Energy Forms

ENERGY FORMS

Allegory and Science in the Era of Classical Thermodynamics

Bruce Clarke

Ann Arbor

THE UNIVERSITY OF MICHIGAN PRESS

Published in the United States of America by
The University of Michigan Press
Manufactured in the United States of America
⊗ Printed on acid-free paper

2004 2003 2002 2001 4 3 2 1

A CIP catalog record for this book is available from the British Library.

Library of Congress Cataloging-in-Publication Data

Clarke, Bruce, 1950–
Energy forms : allegory and science in the era of classical thermodynamics / Bruce Clarke.
p. cm.
Includes bibliographical references and index.
ISBN 0-472-11174-4 (acid-free paper)
1. Thermodynamics. 2. Allegory. 3. Literature and science. I. Title.
QC311.2 .C488 2001
809'.9336—dc21 2001000387

For my father

Acknowledgments

I have had the support and helpful advice of many friends and colleagues. Bob Markley and Bill Rossi encouraged me to attend the 1989 meeting of the Society for Literature and Science. At that gathering Donna Haraway gave an astounding lecture out of her *Primate Visions* material. Begun in 1993, *Energy Forms* has been developed from the start within the support structures of the SLS and through a series of conference papers delivered at its annual (and now international) meetings.

Jim Paxson accepted my paper for a MLA Special Session on allegory at the 1993 meeting, thus initiating a warm collaboration on matters allegorical. I encountered Linda Dalrymple Henderson at the Boston meeting of the SLS, where I gave a paper on Edward Carpenter. This semiobscure figure from the annals of premodernism happened to connect my researches to hers, and soon thereafter Linda showed me the portal to the fourth dimension of space, which plane I have been exploring ever since.

I am grateful for a Texas Tech University Faculty Development Leave and a Gloria Lyerla Memorial Research Travel Grant, and for my Tech colleagues who have supported the Center for the Interaction of the Arts and Sciences: Madonne Miner, Jane Winer, David Knaff, John Morrow, Ron Rainger, David Weinberg, Joe Bilello, Mark Stoll, John Burns, David Schmidly, and Donald Haragan. Thanks also to Mrs. Iris McCabe and Frank James at the Royal Institution, London; to the South Central Modern Language Association (SCMLA) for a Harry Ransom Humanities Research Center Fellowship; and to the staff of the HRHRC, University of Texas, Austin.

Thanks to the following persons for the chance to offer portions of this work while in progress: Michael Bell and the "Modernism and Mythopoeia" conference at the University of Warwick; Linda Henderson and Richard Shiff at the Center for the Study of Modernism, University of Texas, Austin, and again at the University of Texas symposium "From Energy to Information"; Vicki Hill, John Lewis, and Southern Methodist University for the opportunity to speak at Fort Burgwin; Earl Ingersoll and Keith Cushman on behalf of the "D. H. Lawrence and New Worlds" conference in Taos, New Mexico. As this project was nearing completion, it greatly benefited from dress rehearsals at the General Lewis lecture in the Department of English, Texas A&M University; the Comparative Literature Program, University of South Carolina; and the Writing Science seminar at Stanford University. Thanks to Larry Mitchell, M. Jimmie Killingsworth, Allen Miller, Tim Lenoir, and Steven Meyer.

For aid and assistance I would also like to express my gratitude to Luis Arata, Gillian Beer, Linda Brigham, Jim Bono, Donna Clarke, Jona Clarke, Hugh Crawford, LeAnn Fields, Kate Hayles, Ken Knoespel, William Paulson, Sid Perkowitz, Stuart Peterfreund, Arkady Plotnitsky, Martin Rosenberg, Ron Schleifer, Susan Squier, Steve Weininger, Norton Wise, and Michael Wutz.

Earlier versions of some material in this work have appeared as "A Different Sun: The Allegory of Thermodynamics in D. H. Lawrence," in *Myth and the Making of Modernity,* ed. Michael Bell and Peter Poellner (Amsterdam and Atlanta: Rodopi, 1998), 81–98; "A Scientific Romance: Thermodynamics and the Fourth Dimension in Charles Howard Hinton's 'The Persian King,'" *Weber Studies* 14, no. 1 (winter 1997): 62–74; "Allegories of Victorian Thermodynamics," *Configurations* 4, no. 1 (winter 1996): 67–90; and "Introduction: Allegory and Science," *Configurations* 4, no. 1 (winter 1996): 33–37.

Grateful acknowledgment is made for the following:

Permission to publish an extract from a letter of Herbert Spencer to John Tyndall, courtesy of the Royal Institution of Great Britain.

Quotations from *We* by Yevgeny Zamyatin, translated by Clarence Brown, copyright © 1993 by Clarence Brown. Used by permission of Penguin, a division of Penguin Putnam Inc.

Permission to publish selections from D. H. Lawrence, *Reflections on the Death of a Porcupine and Other Essays* (Cambridge University Press), courtesy of Laurence Pollinger Limited and the Estate of Frieda Lawrence Ravagli.

Every effort has been made to trace the ownership of all copyrighted material and to secure permission for its use.

Contents

Introduction

A vague mist, the distant surf—then a close-up of breaking waves. The shore of the end. Terminal Beach. I must have first read *The Time Machine* when I was ten or eleven, around the time Rod Taylor and Yvette Mimieux starred in the film version. In later years I could always bring to mind this image: a vague mist, far-off surf, then rolling waves on an empty beach. I am Wells's Time Traveller thirty million years into the future, staring thermodynamic entropy in the face. Not that I knew about the classical entropy concept then—just the eerie isolation of that scene. But later, when I learned about the second law of thermodynamics, I had an image to attach to it.

I must have first seen the movie years later, probably on TV, and after that remembered the story through the movie, with no memory of ever reading the book. Then in the 1990s, reworking through my literary field with a new attention to science matters, I wound back to *The Time Machine*, convinced that I was finally getting around to the literary source of that hokey movie. And there toward the end of the text was the sad surf and the dim sun. It made me curious to see the movie again. When I did I caught my own misrecollection. In the movie there is nothing of that—no last leap into the far future, no beach, no sublime terror of cosmic dissipation. Only, as the Time Traveller is first starting out toward the future, the brief interpolated 1960s moment of sirens wailing nuclear attack.

In these displacements of memory, a cold-war liberal-Protestant Anglo-Saxon American adolescent's response to a fin-de-siècle British scientific apocalypse was repressed and reinvented, rescreened by further anxieties about science and technology, politics, conflict, and extinction. That far-off surf, the lonely sea at the end of *The Time Machine*, I will argue, is one prominent example of the social and moral scientisms produced when early thermodynamics mixed with scriptural and cultural prophecies. This study of the cultural poetics of energy forms in the era of classical thermodynamics unfolds from and perhaps fulfills some cluster of these images absorbed in childhood and adolescence.

In *Chaos Bound: Orderly Disorder in Contemporary Literature and Science,* Katherine Hayles notes that "the older resonances do not disappear" when "a word such as 'chaos,' invested with a rich tradition of mythic and literary significance, is appropriated by the sciences and given a more specialized meaning."[1] This study begins by examining another and perhaps even more far-reaching semantic appropriation that occurred in mid-nineteenth-century science. In the 1850s, an earlier science of physics borrowed the word *energy* from a traditional register of literary and philosophical significances. Then as now, the authors of scientific discourses reached into the archives of mythology and literature, not just for rhetorical embellishments, but also for the terms that named new conceptions of natural objects and processes. Untidy and promiscuous cultural significations enter the mansion of science through the back alley, like Mr. Hyde, and then step out through the front door, like Dr. Jekyll, in fresh lab coats. Science and the humanities often keep their consubstantial identities and their transformations into each other from public view. *Energy Forms* seeks to repair some of these self-divisions of modernity by examining significant episodes in the discursive commerce between literature and science—in particular, allegory and physics—in the later nineteenth and earlier twentieth centuries.

Energy was transferred to physics from its theological and humanistic provenance as part of the conceptual reformulations brought about by the new discipline of thermodynamics. But in this case as well, its prior layers of meaning did not vanish. The already overdetermined term *energy* became even more charged with powerful semantic currents. Emotional and spiritual meanings were mingled with the letter and interpretation of physical concepts. This discursive overlapping significantly affected the cultural reception and social elaboration of the new scientific laws of energy. To assess these effects, this study reviews the history of physics from the inception of modern energy laws to the incipient dissolution of the late-classical paradigm with the emergence of atomic physics, quantum mechanics, and the special and general theories of relativity. We will follow the modern cultural development of energy concepts, both as they enter scientific discourse from earlier philosophical sources and as they then reenter the literary and social arena augmented by the detour through physics.

The onset of the era of classical thermodynamics is easier to determine than its end. Within the school of British physics at the center of this study, it is convenient to date the arrival of classical thermodynamics with the publication of William Thomson's paper "On a Universal Tendency in Nature to the Dissipation of Mechanical Energy" in 1852. Although many others—James Joule, Robert Mayer, Macquorn Rankine, Hermann von Helmholtz,

James Clerk Maxwell, and Rudolf Clausius—share the honor of first formulating thermodynamic principles from the 1840s through the 1860s, in the Anglo-American arena it was Thomson who first articulated both the immediate terms and the cosmological consequences of the new science of energy, the "universal" reach of mechanical dissipation—the rule of entropy stated by the second law of thermodynamics.

According to Thomson, in something like thirty million years (the calendar Wells follows in *The Time Machine*), the solar system is fated to exhaust its energic resources and enter a terminal state of "heat death."[2] The notion of heat death predicted universal demise not from heat per se but from the ultimate conversion of all "higher" energies—such as chemical bonds and electromagnetism—into the "lowest" form of thermal motions. This redistribution of all energies as heat yields a cosmic equilibrium with all temperatures leveled off to a mean of three degrees above absolute zero. According to Thomson's scenario, at this point no further mechanical work can ever be extracted from the universe, "unless operations have been, or are to be performed, which are impossible under the laws to which the known operations going on at present in the material world are subject."[3] The popularization of this unsettling doomsday vision or allegory of last things (fig. 1) as the last word in deterministic scientific predictions created much of the cultural resonance of classical thermodynamic ideas.

The thermodynamics of this period is "classical" not just in being the originary phase of what has since become a complex and evolving history. In addition, the presiding disciplinary paradigm into which thermodynamics was fit at its inception is now commonly called "classical physics"—the Galilean-Newtonian synthesis of mathematical and dynamical principles for operating on matter in motion through space and time. Corollary axioms of classical physics presumed the mechanical determinism of idealized dynamical systems—their orderly predictability—and the material continuity of the cosmos necessary to allow coherent mechanical descriptions of radiant phenomena. This later assumption supported the concept of the ether of space, a hypothetical substance evenly diffused throughout the universe and providing a subtle material medium within which the movements of light, heat, and electromagnetism could be conceived as the propagation of continuous waves. Toward the end of the nineteenth century, crises in the theory of the ether heralded the breakup of late-classical physics. Then in the first decade of the twentieth century, Einstein's theory of special relativity eliminated the need to posit the ether medium as an absolutely stationary cosmic reference frame.

Within most scientific discourse, the ether of space "ebbed quietly out of

Fig. 1. Ce sera la fin. (This will be the end.) From Camille Flammarion, *La Fin du Monde* (1893).

the universe."[4] However, by that time, the hypothesis of heat death as well as the doctrine of the ether had rippled far beyond scientific discourses and gone through a number of cultural transformations, reflected in a broad range of social, philosophical, and aesthetic repercussions. The cultural era of classical thermodynamics did not end, for instance, once the new atomic physics exploded William Thomson's mechanically foreshortened predictions of solar life expectancy, nor once Einstein dismissed the absolute geometry of the ether from a relativized world-picture. With the hardy cultural resilience of scientific anachronisms, they continued circulating within many scientistic progeny. Nevertheless, this narrative draws to a close at the end of the 1920s, on the consideration that by this time the more advanced outposts of informed opinion had registered the consolidation of the paradigm shift from a continuous mechanical ether to the electromagnetic and gravitational field theories from which we now date our current understandings of the discipline.

The critical goal of my historical method is to articulate these temporal disjunctions in sufficient detail to avoid anachronistic misreadings of contemporaneous cultural productions, while at the same accurately reading their anachronisms. I have been led in this direction by art historian Linda Dalrymple Henderson's analysis of the problems in art history that resulted when the discourse of the fourth dimension in cubist and other early-modernist manifestoes was construed, anachronistically, according to the four-dimensional space-time continuum of Einstein's general theory of relativity, rather than, more accurately, according to the spatial fourth dimension constructed in the later nineteenth century out of *n*-dimensional geometry.[5] The formal and philosophical motives of early modernist abstraction, at least in the first two decades of the twentieth century, derived not from the space-time of Einsteinian field theory but from earlier discourses like that of late-classical hyperspace, which emerged alongside classical thermodynamics and its concomitant ether. For an adequate historical contemplation of artistic and discursive works completed between 1850 and 1930, then, one must place contemporary scientific paradigms in proper relation to the axioms and commonplaces of late-classical dynamics.

Together with its protean counterpart and antagonist *evolution,* the forms of energy under classical thermodynamics contributed scientific authority to some problematic strands in the major secular narratives of modernity. During the era of classical thermodynamics, the idea of energy was framed almost exclusively as a closed-system phenomenon seeking a final state of thermal equilibrium. In the period of this particular energetic bias, cultural development was often interpreted as a problem in mechanical engineering. Social progress was often measured as the power to prevent or delay the decline and dissipation of energy resources. Life could also be reduced to a manifestation and maintenance of energy forms and subjected in its entirety to the material economies of physical systems.[6] The assigning of cultural authority to these dominant positivist scientisms generated in their turn vehement, often visionary counternarratives and critical responses, and these significantly shaped the creative literature and imagery of the period. *Energy Forms* looks at a series of texts, some that construct and some that critique the comprehensive natural, social, and cosmological systems propelled by scientific metaphors and images produced in the discourses of classical thermodynamics and early electromagnetics.

Part 1 develops three interrelated modes of critical approach to the literary and scientific history under discussion. The central thread in this discussion is the theory of allegory. Chapter 1 argues that the history of Western allegory provides a conceptually efficient frame within which to connect the discursive relations of literature and science, and introduces several of the key texts to be studied as exemplary instances of the creative implosion of allegory and science. Much has been written about the interplay of metaphor, rhetoric, and science.[7] I have tried to contribute to these discussions by showing that the specific formal characteristics of allegorical construction allow for a mobile synthesis of otherwise dispersed and disconnected elements. Contemporary allegory theory graphs with great power and clarification the comprehensive lines of connection among the figures and concepts that circulate in and out of science and culture.[8] Tellingly, the term *energy* itself is a powerful allegorical operator: cited in Aristotle as a function of rhetorical power, its prescientific provenance helped to mediate a new semantic interchange of physical and metaphysical images and doctrines. Similarly, the spatiotemporal features of allegorical construction given classical expression in Plato's parable of the cave, a seminal interplay of mathematics, philosophy, literature, and science, reach forward to the cultural milieu of classical thermodynamics. The parable of the cave is a virtual subtext for the spatialized thematics of scientific knowledge and cultural ignorance in Victorian dimensional allegories such as Bulwer-Lytton's *The Coming Race* (1871), Edwin Abbott's *Flatland: A Romance of Many Dimensions* (1884), Charles Howard Hinton's *The Persian King* (1886), and H. G. Wells's *The Time Machine* (1895). Traces of the cave will also be significantly discernible in the two literary authors this study positions as lightning rods for the modernist reception of and allegorical response to classical thermodynamics, the Russian novelist and critic Yevgeny Zamyatin and the British writer D. H. Lawrence.

Allegorical forms are intrinsically anachronistic structures that interfold the historical past, the perceived present, and the imagined future. Chapter 2 approaches the collusion of allegory and science in its historical and ideological dimensions by way of another major modernist thinker, Walter Benjamin. Benjamin revived the theory of allegory in *The Origin of German Tragic Drama,* completed in 1925, in which the emblematical resources of baroque dramaturgy received a seminal philosophical rehearsal.[9] He then extended those insights through an allegorical decoding of Marx and Baudelaire into an inspired critique of industrial modernity. *Energy Forms* appropriates to its analysis of classical thermodynamics Benjamin's insistence on reading allegory's temporal dynamic within a materialist framework of irreversible historical development.[10] Benjamin treats the commodity as a "phantasmagoric" object positioned within a "mythic history" constructed by economic relations, and his Arcades project provides some precision tools for critiquing the socioeconomic allegories embedded in those nineteenth-century popularizations of thermodynamics that purveyed a pervasive commodification of the cosmos.[11] From this perspective, chapter 2 examines the sociopolitical allegory in Bulwer-Lytton's *The Coming Race,* as well as two lively and tendentious popular narratives of energy laws, *The Conservation of Energy* (1874) by British physicist Balfour Stewart, and *La Fin du Monde* (1893) by French astronomer Camille Flammarion, translated into Wells's London as *Omega: The End of the World.*

Chapter 3 completes the construction of theoretical frames in *Energy Forms* by appropriating the sociohistorical term *scientism* for the theory of allegory. In previous critical usage the concept of scientism has generally provided a derogatory category for inappropriate extensions of scientific procedures and prestige into nonscientific activities. However, appropriate or not, scientism is to allegory as the figurative extension of a scientific concept is to the textual extension of a literary metaphor. Recent work in the social and literary history of science by Donna Haraway, Bruno Latour, Katherine Hayles, Greg Myers, Stuart Peterfreund, Gillian Beer, James Bono, and many others has demonstrated that the domain of science is inseparable from both its technological and its wider cultural ramifications. Latour indicates the implosion of science and culture with the term *technoscience,* in which the splicing together of science and technology also implies their inextricable relations to society as a whole.[12] My own neologism *technoscientism* resituates scientistic discourses within the total cultural field of technoscience. It implies further that the structures and processes of allegory have operated and are continuing to operate throughout the modern and postmodern cultural fields to provide formal shape, rhetorical texture, and ideological force to the circulation of scientific and technological discourses.

Chapter 3 details the range of Victorian thermodynamic scientisms, locates the electromagnetic ether of classical thermodynamic technoscientism as a conceptual ghost in the theories of the "cultural field" at large in contemporary critical theory, and then focuses on the interplay of energetics and vitalism that crests in the early twentieth century. The energic ambiance of early-modernist vitalisms is noted in the social and political discourses of visionary scientistic commentators Edward Carpenter and Dora Marsden.[13] The self-definition of modernist vitalism against the classical discourse of physical energy is then traced in a scene from D. H. Lawrence's *The Rainbow,* in which the protagonist, Ursula Brangwen, establishes her own existential absoluteness through a vitalistic defiance of mechanistic reduction. As a creative counterscientism, Ursula's vitalism is related to Zamyatin's own defensive conception of an energic recuperation of social entropy in which the first law salvages the second, at the price of interminable upheavals and endless revolutions.

A particular set of technoscientistic formations distinguishes the era of classical thermodynamics from preceding and following periods. In the preceding period, Newtonian scientisms had championed for broad cultural consumption and adaptation the standards embodied in the marriage of physics and the calculus: predictability, determinability, reversibility. This conceptual regime analogized all systems to the frictionless dynamics of solid celestial bodies, modeled ultimately on a clockwork vision of cosmic motions in accordance with gravitational laws.[14] An emblem or allegorical icon of high Newtonian scientism is Laplace's Demon, a perfect intelligence whose universal omniscience derives not from any unfathomable divinity but from the empirical knowledge of the position and velocity of all material particles and the rational capacity to perform accurate calculations upon them. Classical thermodynamics attempted to fit itself into this classical-dynamical regime but was quickly bedeviled by the problem that heat flow is a *fluid* rather than solid phenomenon, the equations for which are greatly less tractable than those for the orbits of planets or trajectories of projectiles. Even less solid and substantial are the phenomena of other radiant energies such as light and electromagnetism. Gradually the conceptual regime of thermodynamics has come to accept a large portion of unpredictability, indeterminacy, and irreversibility. But not without a struggle. The current regime of far-from-equilibrium thermodynamics, which accepts and exploits the local interplay of chaotic and orderly dynamics, emerged only after a long period of conceptual antagonism toward the fundamentally discontinuous and relativistic nature of energic phenomena.[15]

As a result there is a persistent tension in the field of classical thermodynamic scientisms. The unresolved opposition between the residual authority of Newtonian physics and the observational intimations of dynamical chaos is registered in the anachronistic complexes of classical energic allegories. On the one hand, in its Newtonian mode, at first thermodynamics extended the second law to produce the technoscientistic prophecy of universal heat death, the unexpectedly dismal final horizon of the clairvoyance possessed by Laplace's Demon. On the other hand, in its incipient chaotic mode, early thermodynamic scientisms extended the first law into equally imaginative scenarios of the ultimate conservation and recovery of cosmic energies. The foremost icon of this conceptual ambivalence in classical thermodynamic scientism is Maxwell's Demon, a fictive intelligence straddling the seeming disjunction between the first and second laws, conservation and dissipation. Maxwell's Demon opens up closed systems on the inside, reversing the time of entropic dissipation by reestablishing the spatial separation of thermal motions. Thus Maxwell's Demon looks backward to the temporal reversibility of Newtonian dynamics as well as forward to the systematic possibilities of unpredictable novelty and the creative interruption of deterministic processes.

During this period, however, the self-organizing potential of dynamical systems could find little place in physics proper. It was thus displaced to another great scientistic tradition, the counter-Newtonian discourse of vitalism. The doctrine of vitalism provided a conceptual repository for intellectual and spiritual convictions concerning the nonmechanical and incalculable nature of living beings. In the 1850s, moreover, just as thermodynamic ideas were being codified into physical laws, Darwin began to publish his theories of evolution, with their own particular notions of the irreversibility of evolutionary time and the randomly unpredictable genetic swerves that drove evolutionary processes. Thermodynamic and evolutionary technoscientisms twined around each other from the start. To track the development of biological scientisms such as social Darwinism and evolutionary vitalism would take this study too far afield. However, the social and literary history of evolutionary theory comes within the purview of *Energy Forms* when a classical vitalistic concept like the *Lebenskraft,* or life force, fully functioning since the eighteenth century, is refashioned in the mid–nineteenth century on the twin models of evolutionary progression and physical energy. In a particular late-classical energy form I term *electrovitalism,* the aura of solar and electromagnetic power and propagation is transferred from physical to physiological and existential applications. Electrovitalisms shadowed energic and entropic scientisms throughout the era of classical thermodynamics and were especially prominent in the modernist period.[16]

Part 2 focuses closely on the cultural allegory of the first and second laws of thermodynamics. Chapter 4, "Maxwell's Daemonic Science," examines the explicit allegorical ambiance in the personal and professional writings of the remarkable British physicist James Clerk Maxwell. The cameo roles Maxwell and his famous Demon perform in Thomas Pynchon's *Crying of Lot 49* introduce the abiding literary resonance of Maxwell's heuristic thought-constructions. Maxwell himself was a modestly accomplished author of satirical verse, and his poetic works indicate the tenor of his allegorical culture. However, it is in his scientific papers that that culture comes to astounding fruition. Maxwell's scientific writings, especially those in which he develops Faraday's insights into his own seminal theories of the electromagnetic field, are quite explicit adaptations of allegorical methods to the mechanical modeling of physical systems. His literary bent and heuristic panache were then memorialized in the imagination of a "neat-fingered being," christened by his mentor and colleague Lord Kelvin as Maxwell's Demon, created in the context of a thought-experiment designed to "pick a hole" in the second law of thermodynamics. Seldom have allegory and science come together in more epochal fashion than when Maxwell indulged himself in the daemonic construction of a divided microcosm inhabited by a sorting-demon. But it was his colleagues Thomson and Tait who then put the Demon on the cultural map, by appropriating its figure to their own popular moralizations of thermodynamic regimes. By then, despite Maxwell's own reluctance to own the creature, the genie was out of the bottle. The Demon has played a significant role in thermodynamic and informatic theorizing ever since.

Chapter 5, "Dark Stars," further develops the allegorical implications and repercussions of the concept of entropy. Just as religious allegory had been a standard vehicle of apocalyptic prophecy, scientific allegory comes forward in this period to convey the terminal vision of universal entropy. Theology and science approach particularly closely on the discursive field of cosmology: the second law had such a profound effect on modern culture precisely because of its propensity to cosmological extension. Moreover, the British thermodynamicists, including Maxwell, were adherents of a Christianized form of natural philosophy seeking to reconcile Newtonian mechanism and a supreme deity. Consequently, their popular physical expositions were surcharged with the forms of spiritual allegory. However, the era of classical thermodynamics was also one of great religious eclecticism. The visionary scientisms that developed beyond the early theologizing of thermodynamics were variously idealistic, theosophical, spiritualistic, or secular and positivistic, depending on their particular authors.

Charles Howard Hinton's energic fable *The Persian King* seeks moral asylum from the materialism of the second law in the conceptual haven of the spatial fourth dimension. In this process Hinton developed a remarkable anticipation of current revaluations of thermodynamic entropy. H. G. Wells's socioscientific parable *The Time Machine* co-opts Hinton's fourth dimension and refashions it as a spatialized medium of time transportation. But the ultimate destination of the Time Traveller's pilgrimage brings neither him nor the world any spiritual or material recuperation: he dead-ends on Terminal Beach at the brink of solar exhaustion and cosmic heat death. Two years later, in the lesser-known short story "The Star" (1897), Wells imagined the more immediate and cataclysmic catastrophe of a comet's near-collision with the earth. In a tidy reworking of Camille Flammarion's full-length didactic allegory of scientific prophecies in *La Fin du Monde,*

"The Star" sketchily suggests a postcatastrophic renaissance of human society. In Flammarion's text, medieval allegoricism is adapted to a prominent form of scientistic spiritualism. At the end of our world, higher evolution is spared the cosmic insult of heat death by a kind of ethereal apotheosis of the last human remnant. Flammarion's final apocalypse marshals the conservation of energy and the force of gravity against the second law to guarantee the universe a vital and infinite future run.

The modernist period was handed this dubious inheritance of unresolved tensions, both within thermodynamic ideas and generally within technoscientific culture. Chapter 6, "A Different Sun," compares the complex clash of positivist and visionary scientisms in the writings of Yevgeny Zamyatin and D. H. Lawrence. In their respective milieus, Zamyatin and Lawrence occupied the cusp of the transformation from classical to contemporary physics, and both inscribed their writings with emotional and ideological fissures connected to this particularly charged moment of political, intellectual, and cultural revolution. Zamyatin's great novel *We* is literally fashioned out of allegorical dramatizations of warring scientisms. Along with antithetical elements of good and bad mathematicisms and evolutionisms, variously polarized thermodynamic rationales are parceled out to the main personae. In the process Zamyatin satirized contemporary representatives of cultural energetics—implicitly, in the case of Wilhelm Ostwald, and explicitly, in the case of Frederick Winslow Taylor. His characters seek psychological and social regeneration in sexual vitalism and revolutionary energy. Yet despite its prescient critique of Soviet and Fascist totalitarianism, Zamyatin's fictive dystopia is self-condemned to captivity at the thermodynamic impasse between the order of crystalline regimentation and the chaos of random thermal equilibrium. *We* remains locked in the totalizing cell of classical thermodynamic closure.

Lawrence also saw the classical thermodynamic cosmos as a space of spiritual confinement, an imprisoning womb, and hoped to reestablish a vital relation between himself, human beings, and the cosmos, under a "different sun." But many of his escape routes are similarly impeded by the modernist impasse between order and chaos. His doctrinal writings develop into a deliberate fantasia in which a phantasmagoric sun is placed into a pervasive circuit of reversible cosmic polarities. He was powerfully motivated in this direction by the technoscientistic density of the cultural dilemmas he was addressing. But when Lawrence elaborated his electrovitalist schemes directly under the sign of thermodynamic anxieties, he often fell into other versions of the closed systematics that he was trying to overcome.

In allegories of classical thermodynamics, the apocalyptic finale arrives as cataclysm or inertia, a bang or a whimper. Neither of these versions of scientized apocalypse is the last word.

Part 3 shifts energy forms from thermal dynamics to electromagnetic radiation. Chapter 7, "The Ether of Space," examines what Sir Oliver Lodge called the "many effects" of the luminiferous ether in late-classical wave theory. The ether was conceptually bound up with the construction of classical thermodynamics, but it was also the site of some pressing theoretical difficulties. Formally considered, the geometrical and conceptual matrix of the ether was an allegorical construction of the first magnitude, one that became elaborated into its own ensemble of fantasias. But the ether also showed the way beyond phantasmagoric scenarios of entropic dissipation. The revisions of this scientific fiction of radiant propagations yielded the current shape of physical space-time.

Allegory and science were deeply interpenetrated in the many effects of the many ethers spun off from Victorian physics and metaphysics. For instance, British geophysicist John Tyndall celebrated the ether as a sublime entity that accommodated both reason and imagination. Classical thermodynamicists Balfour Stewart and P. G. Tait posited the ether as the medium of energic connection between the material world and an Unseen Universe. Theosophists heralded the ether's scientific validation as a confirmation of the circulation of thought waves upon astral planes. Supporters of the concept of the fourth dimension of space used the electromagnetic dynamics of the luminiferous ether to prop up the physical plausibility of their geometric thought-constructions. The ether and the spatial fourth dimension came forward to early modernism as full-blown and mutually reinforcing mathematical and physical scientisms. Especially for the well-populated transcendental wing of modernist art and literature, these pre-Einsteinian allegories of space and time provided dynamic models for the aesthetic transition from mimesis to abstraction. The occulted connectivity of the ether in these developments keeps modernist abstraction within the field of its original cosmological rationales. Later notions of flat abstract surfaces mark the forgetting as well as the abandonment of the concepts of hyperbolic depth in late-classical physics and early modernism. The allegory of ethereal space and time is recognizable now as the sociospiritual analogue of the development of electronic communication technology.[17] As the prior era of classical thermodynamics wound down, the luminiferous ether-medium was sublimated into electromagnetic fields and wireless media.

The shifting play of time frames within allegory aligns it with the mathematical reformulations of time within postclassical physics. Chapter 8, "Literary Relativity," tracks the lapse of the ether of space against the rise of the Einstein phenomenon and the advent of four-dimensional space-time. The new concepts of physical relativity were broadly embraced in the years following the Great War, when the paradigms as well as the palaces of the old world had been shaken or broken by intellectual and technological revolutions. If we now see Einstein as the last great physical classicist, in certain ways still wedded to idealized forms of continuity, totality, and reversibility, it is because we have been raised in a later scientistic milieu of generalized relativity and chaos. But modernist culture perceived Einstein's moment—the moment when modern thought broke substantially as well as notionally from the absolutist foundationalism of earlier science and philosophy—as a momentous one, and his theories of relativity sent waves through the cultural field that initiated much rethinking of cosmological positions and rewriting of scientistic allegories.

In 1920, Edward Carpenter was moved to revise his essay from 1889, "Modern Science: A Criticism," in light of the Einstein vogue, but in ways that exposed the ideological fissures of his earlier visionary scientisms. Yevgeny Zamyatin, already heavily invested in classical thermodynamicism, fitted Einstein into his tense parables of heretical rebellion against dogmatic positivism. His writings nicely demonstrate the ease with which the Einsteinian fourth dimension could be fitted to the semantic and connotative frames constructed by the earlier fourth dimension of space. D. H. Lawrence's encounter with Einstein produced none of the smoke and flame of his agon with Freud. Lawrence was not averse to manipulating the equivocations of "relativity" in the direction of his own thematics of existential relation. The Einstein phenomenon also appears to have catalyzed for Lawrence a certain resolution of the thermodynamic impasses that register elsewhere in his text. His later writings expound a creative fourth dimension from which the painful relativities of space and time are momentarily absolved. Lawrence's absolutist relativity provides an attractive mode of positive anachronism, a visionary scientism in which entropic time and ethereal space yield a resilient allegory of perpetual self-recreation through other-relation.

In the historical and technological arc from steam engines to solar panels, from locomotives to space vehicles, from heavy industry to communications technology, thermodynamic entropy is not eliminated. But in the immaterial mode of radiant energies, its conceptual oppressiveness is lightened. In the physics of field theory it can be "black-boxed" or bracketed more readily, while remaining as one dimension of the system. Coming into our current thermodynamic era, cybernetics and information theory, dynamical systems theory, and the science of dissipative structures have relativized the concept of entropy altogether and resolved to a great extent, at least conceptually, the antagonistic impasse between order and chaos. In the period preceding ours, these developments tended to cluster around the physical conundrums of heat and electromagnetism. The concept and technoscience of *information* is now the premier recipient of a mobile fund of cultural cachet previously invested in the energy concept. In the era of classical thermodynamics, the cultural allegory of energy forms proceeded from the universal moralization of heat to the relativistic decoding of light, preparing both matter and energy for further metamorphoses into the chaotic orders of information.

PART 1

FORM, IDEOLOGY, SOCIAL HISTORY

Chapter 1
Allegory and Science

Models of Energy

In *Hermes: Literature, Science, Philosophy* (1982), which brought representative samples of Michel Serres's earlier works to an English-speaking audience, the chapter "Turner Translates Carnot" famously connects energy forms and artistic styles. Serres interpreted works such as *The Burning of the Houses of Parliament* and *Rain, Steam, and Speed: The Great Western Railway* as visual allegories for the nineteenth-century onset of thermodynamic principles. Turner's expansive pictorial combinations of sun and ocean symbolize the industrial as well as cosmic generation of "mechanical effect" or kinetic energy provided by the descent of heat from hot to cold bodies. In previous eras, industrial power had been seized from modest forces linked to simple machines—wind and sails, waterwheels, mills, and muscles. In the transition to a culture sustained by steam technology, according to Serres, this world of pulleys and harnesses is consumed in the chaotic fury of heat engines.[1] At the threshold of the era of classical thermodynamics, Turner's paintings forecast a period when a significant portion of artistic and discursive innovation would be driven by an accelerating culture of physical energies.

The forms of physical energy developed in the sciences of thermodynamics and electromagnetism and deployed in the colossal technologies of industrial modernity have lent themselves to a broad expanse of allegorical production. To mediate and moralize the relations of different places and eras, disparate realms of activity, and different levels of being, stories and images deployed new ensembles of physical concepts and fictional elements.[2] Narratives and pictures that registered and inscribed the new orders of energy rendered these physical renovations of time and space into allegorical structures through interrelated figures of rhetorical and physical agency. At the same time, these and other artistic and discursive works also transformed the cultural space of subsequent scientific developments.

Scientific models embody imponderable concepts in tangible forms.

Modern physics gained conceptual and technological purchase on the imponderable forms and phenomena of heat, light, gravity, and electromagnetism not by seizing reality bare-handed but, to a significant extent, through *scientific* allegories, that is, by constructing and investigating as factual fictions increasingly workable models of energy.[3] The critical reading or interpretation of a narrative or a painting as an allegorical structure parallels the modeling activity of scientific theorizing: the art object is not simply a lure for personal identifications but can be approached in terms of a formal calculus and a set of transcoding rules. The allegorical depiction of a cosmos or encoding of a text both constructs a series of literal or figurative equivalences and provides or implies multiple frameworks within which to read those systems of signs. The application of allegorical models across scientific, imaginative, and critical constructions has generated a cultural network of formal and thematic resonances linking the discourses of literature, art, and science within a common matrix. The concept of energy is an especially crucial relay point within this network of physical and social ideas, natural and ideal concepts.

The era of classical thermodynamics begins in earnest in the mid–nineteenth century with the emergence of the modern physical concept of energy. This newly scientific sense of the perennial term *energy* was established by a north British group of natural philosophers—James Prescott Joule, William Thomson, Macquorn Rankine, James Clerk Maxwell, and Peter Guthrie Tait—in order to supersede the older notion of mechanical force and, in the process, to secure broader cultural authority by creating a "science of energy" in which physical principles were coordinated with a particular complex of religious doctrines.[4] As Clerk Maxwell explained to one of his mentors and a luminary in the previous generation of British physics, Michael Faraday, "Force is the tendency of a body to pass from one place to another," whereas "Energy is the power a thing has of doing work arising either from its own motion or from the 'tension' subsisting between it and other things."[5] In contrast to the material momentum and impacts of masses under mechanical forces, energy was now conceptualized as a multifarious *power to do work,* a universally conserved and multiply convertible activity or potentiality. In terms of this new energy concept, for instance, one could now speak of the *potential* energy of a chemical bond or of a mass at rest within a gravitational field, as well as the *kinetic* energy possessed by an object due to its thermal or dynamical status. The concept of energy thus unified physical investigation, reoriented chemical and biological research toward the centrality of physical principles, and increasingly came to be regarded as a foundation upon which to integrate all natural processes and cultural activities.

The historical vagaries of the terms *force* and *energy,* as well as the imperial ambitions of those scientists rallying to the new energy concept, inform one of Maxwell's curious verse epistles, a bit of professional doggerel inspired by a paper delivered at the British Association for the Advancement of Science in 1876.[6] Maxwell and his colleagues were at the forefront of the effort to restrict the usage of the traditional term force in favor of the more comprehensive energy concept. In his "Report on Tait's Lecture on Force:—B.A., 1876," Maxwell satirized this "considerable confusion of nomenclature"[7] and supported his friend Tait's desire to renovate the vocabulary of physics with the thermodynamic redefinition of energy. Concerning savants who continued to expound outmoded paradigms of force, the poem declares:

That small word "Force," they make a barber's block,
  Ready to put on
Meanings most strange and various, fit to shock
  Pupils of Newton. . . .
The phrases of last century in this
  Linger to play tricks—
Vis viva and *Vis Mortua* and *Vis*
  Acceleratrix:—
Those long-nebbed words that to our text books still
  Cling by their titles,
And from them creep, as entozoa will,
  Into our vitals.
But see! Tait writes in lucid symbols clear
  One small equation;
And Force becomes of Energy a mere
  Space-variation.

In the new math of energy physics, the concept of force is preserved, but only as a kinetic parameter or "space-variation" of bodies in motion from one point to another. In the new scientific nomenclature: "the spatial integral of force is known as the 'work' done by the force in moving from *A* to *B,* and $\frac{1}{2}mv^2$, previously called the *vis-viva,* is the 'kinetic energy' acquired by the system."[8] Maxwell parodied the sublime tones of Shelley's *Prometheus Unbound* to sing the mock-millennial transformation of Newton's laws of motion into the updated axioms of energy:

Force, then, is Force, but mark you! Not a thing,
  Only a Vector;
Thy barbèd arrows now have lost their sting,
  Impotent spectre!
Thy reign, O force! is over. Now no more

Heed we thine action;
Repulsion leaves us where we were before,
So does attraction.
Both Action and Reaction now are gone.
Just ere they vanished,
Stress joined their hands in peace, and made them one;
Then they were banished. . . .

In Maxwell's playful poetizing of these physical developments, a number of traditional mechanical concepts are personified as captives of the new conceptual regime that has unified them at a higher analytical level through the energy concept. Yet his satirical rhetoric here was a model of subtlety compared to the following excerpt from an anonymous 1884 review of thermodynamics pioneer James Joule's *Scientific Papers:* "Not Copernicus and Galilei, when they abolished the Ptolemaic system; not Newton, when he annihilated the Cartesian vortices; not Young and Fresnel, when they exploded the Corpuscular Theory; not Faraday and Clerk-Maxwell, in their splendid victory over *Actio in distans*—more thoroughly shattered a malignant and dangerous heresy, than did Joule when he overthrew the baleful giant FORCE, and firmly established, by lawful means, the beneficent rule of the rightful monarch, ENERGY!"[9] Rhetorical transposition of inanimate processes into reified personages is typical of textual and visual allegory. Maxwell's poem and the *Philosophical Magazine*'s enthusiastic commentator offer two examples of the ease with which allegorical devices mediate physical concepts and their wider cultural implications.

The abstract analogical foundation of this interplay between allegory and energy physics is the shared quality of *transformation,* the possibility of semantic as well as material and conceptual metamorphoses. Energy assumed its modern scientific meaning in order to encompass the broad phenomenology of physical forms unified by the principle of the *conservation of energy,* also known as the first law of thermodynamics. Numerous energies are constantly being converted from one form to another, altering quality without loss of absolute quantity. Maxwell himself observed that "the principle of the conservation of energy . . . indicates that in the study of any new phenomenon our first enquiry must be, How can this phenomenon be explained as a transformation of energy? What is the original form of the energy? What is its final form? and What are the conditions of transformation?"[10] The first law provides a conceptual framework for an order of physical morphology that operates at one or more removes from the substantial level of material phenomena. This newly conceptualized metamorphic potential within nature was named with a word whose semantic resources are closely allied to the textual transformations of allegory.

Long before its newly stipulated physical sense, *energy* was a common literary and philosophical term with an honorable Greco-Roman pedigree. As a result, the term entered its modern scientific usages carrying a large freight of classical associations. Since antiquity it has named emotional and textual as well as physical intensity, intellectual as well as bodily vigor, and in particular, rhetorical force. The first definition of energy cited by the *OED*—"With reference to speech or writing: Force or vigor of expression"—is annotated with a reference to Aristotle.[11] Assembled a century before the physical makeover of the term, Samuel Johnson's *Dictionary* (4th ed., 1773) indicates the staying power of *energy*'s textual and rhetorical provenance: "Strength of expression; force of signification; spirit; life. . . . 'Swift and ready, and familiar communication is made by speech; and, when animated by elocution, it acquires a greater life and energy, ravishing and captivating the hearers.' *Holder.*"

Energy has also meant power held in reserve or made actual, brilliant efficacy in function. The possible equivocations, the multivalent expressions, of such power have wired into Western languages a proverbial connection between physics and metaphysics, material effect and divine cause. The *OED* cites Henry More's *Song of the Soul* from 1642: "*Energie* . . . is the operation, efflux or activity of any being: as the light of the Sunne is the energie of the Sunne, and every phantasm of the soul is the energie of the soul." Johnson illustrates this sense with quotations that underscore the theological cast of energy as an attribute of the divine power of material creation:

> What but God!
> Inspiring God! who, boundless spirit all,
> And unremitting energy, pervades,
> Adjusts, sustains, and agitates the whole. *Thomson*

The classical or prescientific energy concept, then, operated ambivalently between physical and spiritual registers. In the allegorical tradition, relations between the mundane and the divine are typically mediated or interfered with by the realm of the daemonic. The cultural as well as orthographical ambivalence of the classical daemonic and the Judeo-Christian demonic reflects the moral oscillation of knowledge and ignorance.[12] The classical *daimon* is typically an agent of spiritual enlightenment, later often degraded as an agent of demonic possession. As a discipline for the production of the sort of knowledge that enables persons to seize powers previously reserved to the agency of the divine, science has often taken on the allegorical attributes of Luciferian enterprise. This convergence of the daemonic and protoscientific status of energy is well captured by the famous lines of Blake's Devil in *The Marriage of Heaven and Hell* (1790–93): "Good is the passive that obeys Reason. Evil is the active springing from Energy. . . . Energy is Eternal Delight."

The Parable of the Cave

Allegory mediates the cultural forces of the mythic and the scientific through the systematic reinscription of verbal and visual signs. Allegory's oscillating alternation of *mythos* and *ratio* reflects the daemonic logic of communication. Messenger figures—angels or daemons, or human adventurers shuttled between two or more worlds—thematize and personify the uncertain play of allegory, the shaky linguistic contracts of possession and capture among conceptually separated spheres of being. Tendentious moral allegories try to bottle up the daemon that produces their doctrinal persuasiveness; they try to reverse or resolve the virtual instability of their own mediated structures—unsuccessfully: as Paul de Man showed us, allegories are inherently self-deconstructive. The network of figures that allows them to operate also gives their agendas away. The angels of allegory are already possessed by the daemons of interpretation.

Allegory and science are both embedded in overlapping textual and historical traditions that discipline their operation as technologies of meaning. They are intersecting developments of classical Greek writing that converged in the efforts of early philosophers to rationalize the discourse of myth. Plato's dialogues applied systematic analogical reasoning to the philosophical domination of archaic cultural texts. Aristotle's treatises shaped Western science in response to the rationalist challenges of Platonic dialectic, and Greek science progressed by extending the abstractive and formulary attitudes of allegorical reading to the measured contemplation of the natural world.[13]

Allegory has always offered science extended metaphors, flexible heuristic frameworks for investigation and reformulation. But it is also the perennial rhetorical vehicle for systematically propounding religious and philosophical doctrines. Hence its smudge of pedagogy and indoctrination, its post-Platonic complicities with patriarchal hierarchies, its promiscuous deference to the highest bid of symbolic capital. The seminal intersection of allegory and Western reason in Greek thought underscores the intellectual as well as social politics at stake in the use of allegory for heuristic and persuasive effect. In earnest or in ridicule, as a medium or as a satire of doctrine, the mode of allegory gravitates to agencies of power, molding as well as mapping the power relations of competing discourses.[14] Thus it is not at all unprecedented for it to emerge as a formal and rhetorical force in the discourse of physical energies.

The godfather of Western allegory populates his dialogues with "myths" that are in fact tightly controlled philosophical allegories reworking archaic mythemes in order to construct rational diagrams for intellectual procedures. To begin one of Plato's most seminal myths, the parable of the cave that opens book 7 of the *Republic,* Socrates seizes his auditor with a dystopian caricature of everyday human life, an infernal image for the dismal abjection of minds content to dwell in the dark womb of fleeting material sensations:

> And now, I said, let me show in a figure how far our nature is enlightened or unenlightened:—Behold! Human beings living in an underground den, which has a mouth open towards the light and reaching all along the den; here they have been from their childhood, and have their legs and necks chained so that they cannot move, and can only see before them, being prevented by the chains from turning round their heads. Above and behind them a fire is blazing at a distance.[15]

Here in a flash is a seminal allegorical tableau of nether and upper realms linked by a mediating passage or wormhole, a kind of umbilical corridor. A flickering subterranean blaze opposes the steady celestial light up above. The fetters upon the prisoners mark the capture and captivation of the mind by the senses. Behind their backs, along a raised wall between them and the fire, effigies of material objects are paraded, throwing upon the wall before them the shadows of these mock-ups, which doubly inauthentic replicas they take for real beings.

According to the allegory, then, sensuous appearances are only the derealized and evanescent tracings of a higher metaphysical reality: "To them, I said, the truth would be literally nothing but the shadows of the images."[16] But what if the geomorphic womb of one's accustomed, enclosed world were suddenly convulsed, cracked open? If one of these prisoners were unchained, shown the specular mechanism of their illusory perceptions, and delivered out of the dark, he or she would "require to grow accustomed to the sight of the upper world. . . . Last of all he will be able to see the sun, and not mere reflections of him in the water, but he will see him in his own proper place, and not in another; and he will contemplate him as he is" (§516). Platonic science is the cult of this epiphany.

The parable of the cave generates its ideal reality (A) through a symmetrical series of specular reversals (table 1). The narrative is constructed rhetorically by binding the given phenomenal world of the senses (B) to an array of allegorical vehicles (C). This supplementation transforms the phenomenal world by mirroring that world's own status within Platonic doctrine as *allegory*—an evanescent "appearance" signifying the abiding metaphysical foundation and subtext of the Forms. Arranged in a continued analogical proportion, the figural relation of B to C represents a deformed reflection of what A is to B. That is to say, the allegorical deformation of the phenomenal world into a benighted cave is a rhetorical reiteration of the doctrinal discrepancy between a reified realm of ideal being and the material world of sensory becoming.[17]

Socrates names this pedagogy precisely as allegorical supplementation: "This entire allegory, I said, you may now append, dear Glaucon, to the previous argument; the prison-house is the world of sight, the light of the fire is the sun, and you will not misapprehend me if you interpret the journey upwards to be the ascent of the soul into the intellectual world."[18] The movement "down" into allegorical debasement is reflected back as the movement "up" from delusion to truth. Departure from the cave is rebirth into a new and higher world. This rhetorical train of allegorical figuration in Plato's text contributes to the historical production of the world of forms, which in its Neoplatonic aftermath will be mythically fixed in the form of abstract doctrinal foundations.[19] Most important for our purposes here, however, is Plato's rhetorical formation of a potent and memorable allegorical architecture that constructs the cosmos in terms of disjunct but bridgeable epistemological and ontological levels. The parable of the cave

TABLE 1. Allegorical Levels in Plato's Parable of the Cave

A Ideal Reality	B Phenomenal Appearance	C Allegorical Fiction
Intelligible realm	Physical world	Cave
Philosophers	Citizens	Prisoners
Intellect	Senses	Chains
The Good	Sun	Fire
Forms	Phenomena	Shadows

prefigures those Western narratives that symbolize spiritual confinement or liberation by spatial or temporal locations in and movements among opposed levels, eras, or dimensions of being.

Entropy as Allegory

In the latter half of the nineteenth century, when the Victorian thermodynamicists began to moralize the physical repercussions of the new energy laws, many of the doctrinal stresses they encountered were still determined by the ontological bifurcations staged in Plato's original theater of philosophical power, as translated into the cosmos of Christian doctrine. Within the Western cosmos, Plato's sun links the eternal effluence of the Good to the boundless providence of God, an infinite source of ultimate power absolved from material and temporal contingencies. For millennia this dynamo of being has centered a cosmic allegory of spiritual energy with proverbial cultural force. But the era of classical thermodynamics challenged the metaphysical provenance of the sun in an entirely new way. In particular, Victorian science would test the sun's mythic status as a fountain of life and pledge of renewal by precipitating its material operation out of ideal immortality and into the historical economies of evolutionary and entropic time. Classical thermodynamics pulled the sun off its throne and put the old orb out to work. In this and other ways, the new laws of energy that reformulated the Newtonian physics of force reconfigured, rivaled, and often surpassed traditional theological and philosophical edicts as authoritative discourses of divine and human truth. New cultural parables emerged, adapted to the altered contingencies of a thermodynamic cosmos.

With a master stroke of scientific nomination, in 1865 German scientist Rudolf Clausius coined the term *entropy* for the increment of disorder in a thermodynamic system, the rate at which the energy of the system is being converted into irrecoverable forms: "as I hold it better to borrow terms for important magnitudes from the ancient languages, so that they may be adopted unchanged in all modern languages, I propose to call the magnitude *S* the *entropy* of the body, from the Greek word τροπη, *transformation.* I have intentionally formed the word *entropy* so as to be as similar as possible to the word *energy;* for the two magnitudes to be denoted by these words are so nearly allied in their physical meanings, that a certain similarity in designation appears to be desirable."[20] Borrowing the root of the term *trope*—the linguistic torsion that transforms literal into figurative usages—the neologism *entropy* carried a kind of mythopoetic dynamism into the physical concepts of energy transformation and dissipation. The metamorphic force of allegory found itself on the inside of matter, in the tenor of energy—its conversion and degradation. In the name of entropy, energic and linguistic transformations became further implicated in each other. The discursive path was now wide open for extensions of thermodynamic ideas throughout the cultural field.[21]

In correspondence with fellow physicist Peter Guthrie Tait, Maxwell applied his light touch to the cultural implications of Clausius's innovation. In 1872 he playfully shifted the scene of entropic disorder from physics to psychology: "As for C[lausius] though I imbibed my $\Theta\Delta^{cs}$ from other sources, I know that he is a prime source. . . . In my spare moments, I mean to take such draughts of Clausiustical Ergon as to place me in that state of disgregation in which one becomes conscious of the increase of the general sum of Entropy."[22] A year later he touched briefly on the notion of "moral entropy" and presciently linked the statistics of entropy to what chaos theory calls the butterfly effect, the infinitesimal threshold of destructive or creative instabilities: "Both Moral & Intellectual Entropy are noble subjects. . . . I do not know much about reversible operations in morals. The science or practice depends chiefly on the existence of singular points in the curve of existence at which influences, physically insensible produce great results."[23] Here again Maxwell deftly elicited the potential moral extensions of physical concepts.

To develop the sense of entropy as cosmological allegory, we can recall that allegorical structures result from the narrative elaboration of metaphors, the systematic extension of conceptual parallels. The temporality of any particular metaphor is instantaneous: "the measure of time for the experience of the symbol is the mystical instant."[24] Metaphors are cross-points of semantic freeplay that catalyze the mind's synchronic elements. In contrast, allegories create a cosmos by exploding metaphorical points into the many dimensions of space and time. The move from metaphor to allegory is from momentary positional juxtaposition to sustained spatiotemporal structure. Allegory adds time to the instantaneous flash of metaphor and so precipitates it out of the literary absolute and into the wear and tear of the world. Entropy is an isomorph of allegory, in that the concept of entropy adds irreversible time to the equations of real physical systems. Entropy has been called "time's arrow" because in real systems time is always observed to move in the direction of increasing thermal equilibrium, the diminishing of the distance heat has to fall. A glass of water warmer than its surroundings and unaffected by another heat-source only gets cooler, never gets any hotter than it already is. The second law of thermodynamics advises us that if it were spontaneously to warm up rather than cool off, time would be going backward.

The heavy-industrial imagination of Victorian thermodynamics also combined the notion of irreversible time with the notion of the material world as a closed energy system, a cosmic machine possessing a finite supply of available power. In his popularization of thermodynamics *The Conservation of Energy* of 1872, British physicist Balfour Stewart made explicit the mechanistic allegory of nature embedded in Victorian science: "We may regard the Universe in the light of a vast physical machine, and our knowledge of it may be conveniently divided into two branches. The one of these embraces what we know regarding the structure of the machine itself, and the other what we know regarding its method of working. . . . A universe composed of atoms, with some sort of medium between them, is to be regarded as the machine, and the laws of energy as the laws of working of this machine."[25] Extending this particular mechanical scenario to its logical conclusion, the sun's prodigal operation could lead only to a state of solar exhaustion—an ultimate, perhaps imminent condition of heat death. In the era of classical thermodynamics, entropy is the daemonic shadow a dying sun casts upon a mechanistic cosmos that leaks energy, suffers friction, runs to waste, winds down to nothing.

Where there is cosmic apocalypse, there is allegory. That is, classical entropy reveals a metaphysical fatality operating to determine the moral outcome of physical events. Thermodynamic entropy became cultural allegory by unveiling a scientifically plausible version of last things. In some circles the second law was interpreted as God's withdrawal from the material universe. Operating in parallel with the theological doctrines of the Fall and the promised end, entropy became *the* cosmic metonymy for a God that has absconded from the physical world.[26] In terms of Walter Benjamin's discussion of baroque allegory, the projected outcome of universal entropy is a scientized form of the melancholy ruin: "Allegories are, in the realm of thoughts, what ruins are in the realm of things."[27] But in this case, the ruin of nature as well as of human history is transposed from time past to the apprehension of a certain future.

Ether and Hyperspace

As in its seminal Platonic applications, allegory typically creates a cosmos, a hypothetical or fictive totality that can be plotted on spatiotemporal axes.[28] An allegory is a written or diagramed configuration of mediated communications in which mixed messages move through nested networks. Mediating discontinuous eras and disparate realms, interweaving the threads of a textual web that would net the world in its mesh, the daemonic fictions of allegory weld a composite cosmos together. In allegorical worlds, lines of communication open up to connect separate levels of being. Humanity discovers its own level to be in contact with other, transcendental or subterranean, realms. As the perennial language of cosmology, allegory typically models a concept of world-space through an articulation of nested structures, universal systems with a montage of ontological levels. The hierarchy of moral evaluation or ethical evolution inscribed in an allegory's temporal dialectic is typically given spatial expression through a complexly polarized world of upper and lower realms or parallel, intercommunicating levels of narration.

Samuel Delany discusses the prevalence of the *paraspace*—"an alternative space, sometimes largely mental, but always materially manifested, that sits beside the real world, and in which language is raised to an extraordinarily lyric level"[29]—in contemporary science fiction. Scott Bukatman has aligned Delany's paraspace with Brian McHale's notion of the postmodern *zone:* "an alien space *within* a familiar space, or *between* two adjacent areas of space where no such 'between' exists."[30] Both of these hyperspatial images, however, operate allegorically—the paraspace is an indeterminate or daemonic surface, a zone that mediates the emblematical levels of alternative dimensions—and both are anticipated by the multileveled diegeses or planes of being in earlier parables of energy and entropy.

In the era of classical thermodynamics, allegories of entropic time dovetailed with dynamical parables of transformed space. The late-classical doctrine of the *ether* was advanced to account for the propagation of forces through space within the conceptual regime of Newton's laws. For a mechanistic cosmos of matter in motion, "action at a distance" was inconceivable without some form of continuous contact, some material plenum whose absolute continuities could bridge the discrete entities of the world and bind them into gravitational and spatiotemporal unity. We will study in detail how concepts of energy and entropy and the wave theory of light as an undulation in the luminiferous ether converged in the 1880s with the concurrent development of non-Euclidean and *n*-dimensional geometries in Charles Howard Hinton's discourse of the fourth dimension of space, and how, two decades later, a related amalgam of energetics and mathematics brought about Einstein's incorporation of the geometrical structure of four-dimensional space-time into his physical theories of relativity.

In Hinton's hands the spatial fourth dimension was an allegorical machine for heuristic shifting between physical fictions and the prophetic intimation of a higher consciousness. Hinton's immediate contemporary Edwin Abbott composed a parallel pedagogical parable in *Flatland: A Romance of Many Dimensions* (1884). Rudy Rucker has noted that "Plato's allegory of the Cave can be seen as prefiguring the concept of Flatland."[31] Abbott's two-dimensional realm is to the cave as Hinton's fourth dimension is to the realm of the forms. Hinton's writings also connect the fourth dimension of space directly to Plato's own projects of allegorical world-building and soul-making. Hinton's parable of classical thermodynamics, *The Persian King,* introduces the character Demiourgos. Alluding to the Demiurge—in the *Timaeus,* the fabricator of the universe—Hinton coined a metonymy of Plato himself, especially the Plato who initiates a line of philosophical discourse in which allegory is the typical vehicle of mathematical as well as moral constructions. Hinton placed his debt to Plato at the beginning of his summation of hyperspace philosophy, *The Fourth Dimension:*

> Plato, in a wonderful allegory, speaks of some men living in such a condition that they were practically reduced to be the denizens of a shadow world. . . . Just as a man liberated from his chains could learn and discover that the world was solid and real, and could go back and tell his bound companions of this greater higher reality, so the philosopher who has been liberated, who has gone into the thought of the ideal world, into the world of ideas greater and more real than the things of sense, can come and tell his fellow men of that which is more true than the visible sun . . . .
>
> Now, I take Plato's suggestion; but literally, not metaphorically. He imagines a world which is lower than this world, in that shadow figures and shadow motions are its constituents; and to it he contrasts the real world. As the real world is to this shadow world, so is the higher world to our world. I accept his analogy. As our world in three dimensions is to a shadow or plane world, so is the higher world to our three-dimensional world. That is, the higher world is four-dimensional; the higher being is, so far as its existence is concerned apart from its qualities, to be sought through the conception of an actual existence spatially higher than that which we realise with our senses.[32]

Hinton candidly declares that he has achieved his ideal metamorphosis of the material cosmos—its vast supplementation by the fourth dimension of space—through the extensive literalization of a metaphor. Giving imaginary substance to an allegorical reification, his thought-project pursues the heuristic extension of proportional series of dimensions—in which the given three-dimensional world is the mean between two imaginary extremes, the second dimension of Flatland and the fourth dimension of Victorian hyperspace—into its manifold of physical and philosophical ramifications. With a strong misreading of the precursor poet, he completes his cosmogony by transuming Plato into his own belated creation. The realm of forms was all along, so Hinton intimates, a model of the fourth dimension of space. Plato stands in Hinton as the Western archetype of the allegorical cosmologist.

Allegorical Construction

Two basic and interlocking forms of allegorical construction are *typology* and *reification.*[33] In typological allegory, a precursor text is positioned behind a later text, providing it a hermeneutic key. Typology concerns the formal analogies of elements located in different texts or of separate levels within the same text. The recognition that a given narrative episode—say, a journey to an unknown place—echoes or alludes to a particular pre-text—a prior or paradigmatic narrative—allows the given text to be inserted into an explanatory frame. For instance, I will be arguing that the parable of the cave is a pivotal subtext in many Victorian and modernist allegories of science.

Alternatively, in reification allegory, the kernel of the text is not a prior narrative but a conceptual or moral abstraction. Traditional allegories, typically concerned with moral abstractions, develop some form of picture language with which to talk about emotional states and spiritual values. Reification concretizes abstractions into things, often as agents of some sort, animated characters or personifications. Reification typically creates daemonic figures—either anthropomorphically, as nonhuman entities are endowed with human forms or traits, or metamorphically, as human beings are transformed into signifiers superseding human or natural functioning.

Hinton and Abbott were both pedagogical allegorists working in a common tradition of mathematical and physical reification. For these self-declared "romancers," the daemonic figures of heuristic beings *more* limited (spatially) than human beings—the zero-dimensional point being, the one-dimensional line being, and the two-dimensional plane being—handily evoked the physical limitations of three-dimensional space and time. These preliminary exercises in subhuman, dimensionally constrained forms prepared the mind for the crucial upshift from three- to four-dimensional space. The literature of dimensional allegories is littered with these anthropomorphized, reified "figures"—animated points and lines, triangles and squares, spheres and cubes. The spaces these characters inhabit in geometrical fables are constructed by a form of extended metaphor, the classical device of "transference by analogy," in which conceptual proportions are derived from compound figurative equations.

For instance, if you posit that Duchamp is to art what Joyce is to literature, then you can say, rhetorically, that "Duchamp is art's Joyce," or "Joyce is literature's Duchamp."[34] Similarly, the geometrical allegorist deploys a kind of dimensional analogism with the following proportion: one dimension is to the next higher dimension as the next lower dimension is to it. The fundamental move of the hyperspace allegorist is to assert that the third dimension is to the fourth what the second is to the third, and so to constitute the spatial "levels" endemic to allegorical narratives by means of nested geometrical dimensions.[35] *Flatland* articulates this structure of analogical dimensionality in A. Square's dream of Pointland, when he imagines that his spiritual guide, A. Sphere, "proceeded to initiate me into mysteries yet higher than those I had witnessed, shewing me how to construct Extra-Solids by the motion of Solids, and Double Extra-Solids by the motion of Extra-Solids, and all 'strictly according to Analogy.' "[36]

Like the formal conversions of energy forms codified by the first law of thermodynamics, transitions from one dimension to another also provide the motives for some spectacular metamorphoses. The metamorphic resources of dimensional analogism are especially evident in G. F. Rodwell's paper "On Space of Four Dimensions" (1873).

> Let us now endeavour to realise the condition of a being living in space of two dimensions. . . . Imagine that a man . . . with our own notions of space of three dimensions, begins to stoop forward and to grow so: his eyes survey less space; he stoops more forward; his body forms angles of 80°, 70°, 60°, 50° in succession, with a horizontal plane. Then he is obliged to go on all-fours, his limbs shorten and are gradually absorbed into the mass of his body; he crawls, he creeps; at length his limbs disappear altogether, and he trails himself along and glides like a serpent, moving in a horizontal plane. . . . Now his body begins to diminish in thickness. . . . Now he is a mere plane, an infinitely thin surface; he occupies space approximately of two dimensions; his eyes are on a line.[37]

Rodwell's retrogressive metamorphosis could have been culled from episodes of transformative devolution in Ovid's *Metamorphoses*.[38] Rodwell's imagery of literary metamorphosis, however, is bound up with the Victorian moralization of human evolution as capable of advance or regression. A deformation from three to two dimensions indicates a devolution of capacities, whereas the opposite movement, from three to four dimensions, marks the spiritual progressivism of hyperspace philosophy. Using such analogical transfers as symbolic prompts, Hinton earnestly hoped to induce a higher metamorphosis of the mind.

Edwin Abbott's literary rendition of Hinton's thesis is more guarded, shrouded in satire, but still purveys the influential use of dimensional tropes to convey unstable periods of brief illumination, moments of altered or "higher" cognition that leave one with no physical evidence by which they can be measured or shared. A. Square experiences what it means to get high and then to crash: "I—who have been in Spaceland, and have had the privilege of understanding for twenty-four hours the meaning of 'height'—even I cannot now comprehend it. . . . To confess the truth, I felt that all that I had seen and heard was in some strange way slipping away from me, like the image of a half-grasped, tantalizing dream."[39] Nor could his fellow Flatlanders comprehend his new doctrines: they promptly immured him in an asylum. Allegories of the fourth dimension play on the structural dynamics of daemonic possession and cognitive aberration, and the incarceration of A. Square draws out the paranoid element in the imagination of hyperspace. As we will see in Hinton's *The Persian King,* the fourth dimension can be an occulted zone from which higher beings pull one's puppet strings. Alternatively, the revelation of the fourth dimension can set its prophet apart from the uncomprehending mass and so induce both paranoia and the persecution it fears. From his padded cell A. Square decries Flatland's legislative prohibitions against unauthorized scientific speculations: "I am mocked at—I, the sole possessor of the truths of Space and of the theory of the introduction of Light from the world of three Dimensions—as if I were the maddest of the mad" (6). Plato had expounded similar ironies in the parable of the cave, warning those who achieve higher philosophical illumination that they would be welcomed back into society not with open gratitude but with defensive incredulity.

Numerous other narratives developing paraspatial allegories of science invite interpretation through this Platonic frame. Among those that we will examine later, Bulwer-Lytton's *The Coming Race* (1871), a conservative or Tory satire of political liberalism in defense of colonial motives, modulates the Platonic parable when the narrator discovers an underground world inhabited by a super- rather than subhuman people. Among the species of Ana, we learn, is the race of the Vril-ya, which has developed higher enlightenment and political preeminence through technological control of an energy form called *vril.* In Abbott's *Flatland* (1884), the Platonic proportions are retained—the Flatlanders are less, not more, enlightened than a visitor from "above"—and dimensionally elaborated. Here the satirical separation of epistemological levels is figured not in layers of geological stratification, but in the distinction of a two-dimensional underspace from a hyperspace of three dimensions. In H. G. Wells's *The Time Machine* (1895), the Time Traveller famously encounters the subterranean dwellings of the

Morlocks and must then unravel the enigma the devolved world of the Eloi poses for the fate of socioscientific progress in the world he left behind.

Again, in Yevgeny Zamyatin's seminal dystopia *We* (1920), upstanding OneState mainstay D-503 finds out that his lucent and regimented world is fissured with paraspaces that shatter all he had known and believed. The revolutionary seductress I-330 opens corridors that lead D-503 through underground passages to a new yet archaic world outside the Green Wall. In another ironic replay of Plato's prisoner released from the cave and dragged into the open light, D-503 blinks in wonder at a different sun: "It wasn't our sun, evenly distributed over the mirrored surface of the sidewalks. This sun was all sharp fragments, alive somehow, constantly leaping spots, that blinded the eyes and made the head spin."[40] *We*'s confrontation of previously disjoined worlds—futures and pasts, upper realms and undergrounds—loops back to the Platonic subtext, while his placing of the literary oscillation of utopia and dystopia into the thermodynamic frames of energy and entropy evokes a Wellsian subtext.[41] By Zamyatin's own day, Einstein's four-dimensional space-time has also begun to warp the strictly spatial schemas of earlier dimensional cosmologies, but the special and general theories of relativity will simply transpose the dissonant mechanisms of allegorical shifting among disparate or imaginary times and places to a higher level of strangeness.

In the era of classical thermodynamics, then, numerous allegories of science recast the moralized geography of Plato's cave in the service of ethereal spaces and evolutionary times. The politics of dominant and subordinate groups were retold as forms of fantastic anthropology. Strangers from alien realms stumble upon various races of higher and lower denizens, in possession of lesser or greater technologies of physical power. Political imperialisms, technological disruptions, and spiritual idealisms were revised and rescreened through fictions of evolutionary variations. Such distinctions were moralized through spatial articulations, the vertical and horizontal disjunctions of allegorical beings. Structures of separation and communication—images of lost underground civilizations, displacements to other times or dimensions, or "space travel" in general—allegorized class and race divisions, gender trouble and sex wars, perennial philosophical and religious dualisms, and modern narratives of polarized societies.

There has been this elliptical orbit in Western thought, an alternating historical current: science has found allegory by turns both attractive and repellent. *Mythos* and *ratio* tend to shear away from implication in the other, and at times succeed in the form of disciplinary or obscurantist purifications. But allegory at its best confounds or subverts these epistemological separatisms. Critical and satirical allegories can exploit the unstable mobility of the mode, strategically calling into question the authority of their own ostensible personae. Allegory shapes the critique as well as the cultivation of doctrinal regimes. Like a sphinx or a cyborg it couples together seemingly incommensurable entities. As their daemonic mediator and angel of desire, allegory is a kind of nomadic double-agent, a mobile supplement and delegate binding the threads of the cultural matrix, keeping myth and mathematics in cultural communication.

Chapter 2
Phantasmagorias of Energy

Cosmos and Commodity

Allegories put obsolescence into question. Allegorical structures are typically in transit to obsolescence—they produce even while they resist the obsolescing of the materials they preserve—and yet good ones remain news. The allegorical tenor of Baroque and other emblems of ruin or superannuation—memento mori such as the broken remains of an archaic statue or a defunct commodity like a 386 desktop computer—arises from this disjointed time-structure, a quasi obsolescence that communicates information about the historical particulars of cultural time.

Allegorical temporality is the discontinuous time of overlapping eras. There is always something anachronistic about allegory. Its generic bric-a-brac results from the clash of distinct temporal phases or cultural eras. Counter to symbolist techniques, which seek to collapse time in instantaneous fusions of multiple meanings, allegory operates by means of temporal montage.[1] The spatial stratification or segmentation of allegorical levels often represents an archaeology of distinct historical periods. In allegory some temporal dissonance, some historical clash of periods generates a layered text intended to neutralize or harmonize that dissonance, to embalm or recuperate an obsolescence. In the modern technoscientific period, the present is always already en route to obsolescence, and the allegorical anachronisms of science fiction typically establish a significant dissonance between images of the present and the future.

For instance, in French astronomer and author Camille Flammarion's popularization of Victorian cosmology, *La Fin du Monde* (1893), a series of displacements into the future frames the author's visionary speculations on the fate of the earth.[2] Anticipating by over a century 1998's *Deep Impact* and *Armageddon,* Flammarion imagined that in the semiadvanced world of the twenty-fifth century, a monster comet is discovered on a collision course with Earth. A scientific colloquium is convened in Paris to debate possible effects, which didactic device allows Flammarion to offer a seminar on later nineteenth-century geological and cosmological evolution, especially as these sciences envisioned possible modes of physical catastrophe bringing life on earth to an end. Once the comet passes without irreparable damage, humanity evolves toward an "apogee" of cultural and spiritual development, a sensuous utopia cut short only by the inexorable processes of geological and solar entropy.

Flammarion's quasi-obsolete prophecies underscore the historical ironies that develop from the modern scientific revision of narrative timescapes. *La Fin du Monde* strove to synthesize scientific positivism and bourgeois spiritualism, only to fracture that synthesis by trying to hold the apocalyptic determinations of classical thermodynamic revelations within a parallel positivist discourse of utopian social prophecy. Over a century later, both the apocalypse and the utopia are semidefunct. Yet in its own day, the text as a whole was already drifting toward doctrinal obsolescence through the stunning anachronisms of the *visual* images set forth to illuminate the discursive content. Packaging the scientific innovations of the last fin de siècle within venerable emblematic codes of medieval, Renaissance, and baroque iconography, this popular text did not leap "back to the future," but forward to the past.

The images of *La Fin du Monde* declare the earnestly allegorical tenor of this composite text. The terrible comet of the twenty-fifth century approaches the earth in the person of a furious and deadly sorceress, her figure scintillating with celestial powers, arms outstretched in a posture of evil intent toward the supine and slumbering Earth (fig. 2). In the far future of a perfected world worn back down by Time, a figure of Melancholy will still brood at sunset over the rubble of Paris and the vanity of human glory (fig. 3). Over the lifeless planet predicted by mechanistic discourses of geological deterioration and solar extinction, Queen Death will reign (fig. 4). In Flammarion's text the visual as well as verbal anachronisms common in many allegorical popularizations of scientific ideas peddle shopworn cultural commodities as the perennial forms of nature.

Benjamin's Dialectical Image

Walter Benjamin's interpretations of modernity turn allegory inside out, in part, by decoding the disjunctions in such mythic images of time. Benjamin's critical allegoresis restores irreversible time to cultural forms, aligning nature with history rather than myth. In *The Dialectics of Seeing,* Susan Buck-Morss reconstructs Benjamin's uncompleted Arcades project as a crit-

Fig. 2. On voyait, dans ces figures allegoriques, la Comète menaçant la Terre endormie. (One sees, in these allegorical figures, the comet threatening the sleeping Earth.) From *La Fin du Monde.*

Fig. 3. Paris, le beau Paris, l'antique et glorieuse cité, n'était plus qu'un amas de ruines. (Paris, beautiful Paris, the ancient and glorious city, was no more than a heap of ruins.) From *La Fin du Monde*.

Fig. 4. La Mort souveraine du monde. (Queen Death rules the world.) From *La Fin du Monde*.

ical system that pivots on the conceptual relations between social myth and critical allegory: "It was the Baroque poets who demonstrated to Benjamin that the 'failed material' of his own historical era could be 'elevated to the position of allegory.' What made this so valuable for a dialectical presentation of modernity was that allegory and myth were 'antithetical.' Indeed, allegory was the 'antidote' to myth."[3]

In the 1850s, the laws of thermodynamics as well as the theory of evolution introduced versions of irreversibility into the temporal description of physical and biological systems. The dissipative transformations of energy and the progressive metamorphoses of species now marked the performance of the world-system over time, marked it as a system in time. Today we view the arrival of irreversible time within thermodynamics as bringing physics closer to the actual workings of cosmological self-organization and far-from-equilibrium systems, thus as an advance over the generalized temporal reversibility that functioned within the regime of Newtonian dynamics and other regimes of scientific materialism. But thermodynamic temporality also came forward as mythic time.[4] Hypothetical thermodynamic causality was often troped as a universal and seemingly imminent mythic fate. Another engraving in *La Fin du Monde* depicts the stark human meaning of this vision of heat death (fig. 5).

Fin-de-siècle thermodynamic prophecies of a universe evacuated of vitality by mechanical dissipation were founded on proverbial Western concepts of temporal linearity, fruition, and summation followed by apocalyptic cessation.[5] Irreversible time emerged in science simultaneously in the historical form of random evolutionary transitions and in mythic or phantasmagoric forms of certain, imminent heat death.[6] For Benjamin, however, the burden and the benefit of irreversible time at the social level is that it determines a new cultural ethic—it makes human history into a matter of fabrication rather than fate. Benjamin's practice depends on seeing nature not as mythic predetermination but a temporal and technological construction, something we make as it makes us.

To critique pernicious interweavings of social mythology and scientific ideology, Benjamin developed a "modern form of emblematics," which Buck-Morss graphs upon two conceptual axes. At the intersection of these thematic coordinates stands the *commodity* in the form of a "dialectical image."[7] Benjamin surveyed the modern cultural field through the scope of the commodity to reveal the material subtexts of mystified metanarratives such as "progress," "social evolution," and "the new." The dialectical images of modern commodities circulate through three interlocking discursive dimensions—myth, nature, and history—divided between axes of con-

Fig. 5. La misérable race humaine périra par le froid. (The pitiful human race will perish from the cold.) From *La Fin du Monde.*

sciousness (waking and dream) and reality (nature in time) and sorted into the "axial fields" of natural history, mythic history, mythic nature, and historical nature. Each field possesses what we can call an *emblem* and a *mode.* Natural history's emblem is the fossil, its mode the trace; mythic history's emblem is the fetish, its mode the phantasmagoria;[8] mythic nature's emblem is the wish image, its mode the symbol; and historical nature's emblem is the ruin, its mode allegory per se. Plotted on these axes, Benjamin's Arcades project explores the cultural dreamwork of the modern ideological dialectic between the cosmos and the commodity. For instance: "How gratings—as allegories—have their place in hell. In the Passage Vivienne, sculptures over the main entrance representing allegories of commerce."[9]

We can use Benjamin's map of the phantasmagoric commodity to read some of the interplay between science and myth in the early cultural elaboration of the laws of thermodynamics. Dialectical images populate influential nineteenth-century discourses that circulated with the authority of science while troping nature precisely as a commodity. Such phantasmagoric descriptions are clearly visible in the early discourses of physical energy. For instance, Balfour Stewart wrote in *The Conservation of Energy,* "Energy may be like the Eastern magicians, of whom we read that they had the power for changing themselves into a variety of forms, but were nevertheless very careful not to disappear altogether. . . . It is necessary to penetrate the various disguises that our magician assumes before we can pretend to explain the principles that actuate him in his transformations."[10] Stewart retails this trope of metamorphic agency to expound the first law of thermodynamics, which asserts that the sum of energy in a physical transaction is conserved throughout its transformations. Whether potential, kinetic, chemical, atomic, mechanical, radiant, or thermal—the qualities of energy alter without loss of absolute quantity. In 1874 Stewart coauthored with another eminent Victorian scientist, mathematical physicist P. G. Tait, *The Unseen Universe or Physical Speculations on a Future State,* which contains further mythic analogues of energy's metamorphoses. "We see that whereas (to our present knowledge, at least) matter is always the same, though it may be masked in various combinations, energy is constantly changing the form in which it presents itself. The one is like the eternal, unchangeable Fate or *Necessitas* of the ancients; the other is Proteus himself in the variety and rapidity of its transformations."[11]

The law of energy conservation comes forth in Stewart and Tait as classical patriarchal ideology. Virile energy is reified in opposition to dull feminine matter as an occulted, potent, and conserved presence that manifests itself in metamorphic guises—from vital sun to luminiferous ether to chemical effect to electromagnetic dynamism to thermal flows to dead pools of entropy—without loss of substance or essence. The Parisian satirist Grandville had already mocked this incipient scientific supernaturalism of cosmic energies in *Un Autre Monde* of 1843. Grandville parodied the scientific sage as a sorcerer and bubble-blowing demiurge (fig. 6). Magician and demon foreshadow the gender and class dichotomies of the Victorian thermodynamic cosmos. The first law infinitely conserves God's divine energies and patriarchal prerogatives. The Devil is in the material and temporal details—entropy, friction, and waste, the demons of the second law.

Fig. 6. "The secret of celestial mechanics: the worlds are soap bubbles blown by an old magician, who fails to notice that a female demon is insufflating love, jealousy and other woes into them." From J. J. Grandville, *Un Autre Monde* (1844).

Energy as Capital

But if every gram of material contains this tremendous energy, why did it go so long unnoticed? The answer is simple enough: so long as none of the energy is given off externally, it cannot be observed. It is as though a man who is fabulously rich should never spend or give away a cent; no one could tell how rich he was.[12]

In this passage Einstein retails an economic metaphor for the interrelation of energy and mass under the relativistic revision of physical conservation laws. His casual usage indicates just how commonplace the rhetorical substitution of money for energy has been. When moralized under the social figures of economic relations, the conservation laws of mass and energy press nature toward the ideological enforcement of conservative political attitudes: nature's administration disallows deficit spending. Rigor in the hard sciences has been modeled on the "fiscal" constraints of natural objects and forces: "The description of any event examined by a physicist or chemist was to be founded on budgets of mass and energy."[13]

Especially in its conservative cultural applications, the early discourse of the first law of thermodynamics produced allegories of *capital* enacting Marx's famous analysis of the metamorphosis of commodities. We note that anything metamorphic, anything bearing transformative possibilities—energy, life, the soul—can substitute for or be substituted by the commodity. Through the "theological niceties" of the commodity, metaphors of money mime the natural as well as the supernatural world.[14] Thus, for instance, in order to expound the difference between potential and kinetic energy, Balfour Stewart developed a remarkable composite allegory of class and capital. Just as one's social level can counteract deficits of personal character, so we are told that (potential) "energy of position" can overcome (kinetic) "energy of motion":

> The former may be compared to money in a bank, or capital, the latter to money which we are in the act of spending; and just as, when we have money in a bank, we can draw it out whenever we want it, so, in the case of energy of position, we can make use of it whenever we please. . . . The former has all the independence of a rich man; the latter, all the obsequiousness of a poor one. If we pursue the analogy a step further, we shall see that the great capitalist, or the man who has acquired a lofty position, is respected because he has the disposal of a great quantity of energy; and that whether he be a nobleman or a sovereign, or a general in command, he is powerful only from having something which enables him to make use of the services of others. When the man of wealth pays a labouring man to work for him, he is in truth converting so much of his energy of position into actual energy.[15]

The homely pedagogy of this passage participates in the wider phantasmagoria of energy that developed as the physical concept was recirculated as an industrial or cultural commodity. Stewart's popular text can be read as a cultural occasion when the intricacies of the concept of energy are put on display for public consumption, and they are presented precisely through an ideological allegory of energy laws that renders nature mythic and social myth natural. The use value of the idea of energy as a tool for doing science has been exchanged for an allegory of class relations, a dialectical image in which a commodified energy concept oscillates in the fields of dream between fetish and wishful symbol.

A suggestive passage from Christoph Asendorf's *Batteries of Life* shows the allegory of capitalized energy operating on a separate but related level of cultural significance. Franz Mesmer's theories of "animal magnetism" had recourse to "the romantic idea of aether, which set the image of medically useful electromagnetic forces."[16] Stewart and his colleagues were representative of several generations of physicists in maintaining an essentially congruent theory of the ether as a physical medium necessary to account for the transmission of radiant energies. What Asendorf suggests about the romantic ether fits directly into the phantasmagoria of commodified energy that develops throughout the nineteenth century: "Marx demystifies the aether forces, this postrevolutionary bourgeois idea of the universal connectedness of all things, by naming the true 'galvano-chemical power' holding society together: money is the unmasked electrochemical aether" (155).

For Stewart, the universal ether is a "subtle medium, or cement, that binds the various parts of the universe together"[17] and provides continuity within the cosmos of the Victorian allegory of energy. The commodified ether assumes a value-form equivalent to the conservation of capital throughout the metamorphoses of currencies and commodities that circulate within polarized social fields. In the process, nature itself is transformed into phantasmagoric myth. Grandville's satirical images capture this commodification of the cosmos. In one, a comet becomes the trail of an opulent gown, the glittering raiment of a crowned goddess charioted beneath the celestial chandeliers of a heavenly ballroom past a throng of minor stellar onlookers. This personified comet is precisely the commodity-on-display projected upon a cosmic panorama of ethereal energies (fig. 7). Another image depicts a modern urban development project. Energy as cosmic capital has funded the construction of the universal ether-bridge. It is erected upon and across the natural planets, themselves improved with modern conveyances, turning the universe itself into an arcade or mall, an artificially illuminated stage for the procession of commodities and wish images (fig. 8).

Grandville's visual allegories critique the transformation of the natural world into something measured entirely by human values. Balfour Stewart's popular text purveys just such a bourgeois metamorphosis of the cosmos. After all, from Stewart's point of view, energy is *useful* only if it occupies a "position of advantage" high enough that its fall can be exploited by conversion into work. Stewart provides a second long figuration of this eco-

Fig. 7. "The passage of a comet across the sky." From *Un Autre Monde.*

Fig. 8. "An interplanetary bridge. Saturn's ring is a balcony." From *Un Autre Monde.*

nomic allegory later in *The Conservation of Energy,* along with a premonition of the modern energy crisis:

> The energy of fuel is derived from the sun's rays; coal [is] the store which Nature has laid up as a species of capital for us, while wood is our precarious yearly income. We are thus at present very much in the position of a young heir, who has only recently come into his estate, and who, not content with the income, is rapidly squandering his realized property. . . . Not only are we spending our capital, but we are spending the most available and valuable part of it. For we are now using the surface coal; but a time will come when this will be exhausted, and we shall be compelled to go deep down for our supplies.[18]

As computed by this economic allegory of mythic nature, the bourgeois value of the sun is that it capitalizes the maintenance of nature's factories. Solar energy deposits an income in the world's bank; that capital is transported from place to place through the subtle and elastic ether-medium of money; it transforms itself into fuel for machines, into the commodities produced by those machines; it wastes itself in the ephemeral luxuries of natural processes; it is taxed into inertia by entropic redistribution.

Masters of Energy: *The Coming Race*

With the initial formulation of thermodynamic laws in the mid–nineteenth century, one of the first effects upon Western culture was to bring the inexhaustible sun of perennial human experience—the sun that symbolized the eternal form of the Good—into conflict with the predictions of scientific materialism. The sun descended from the former, unexamined assumption of its celestial eternity into the present material conditions of increasing entropy, at a time when the idea of entropy portended a merely destructive disordering and dissipation of energies. The sun—the primal marker of human time—was now the object of scientific time, a thermal time that demanded new models of solar energy, new orders of mechanical questioning and of imaginative recuperation.[19]

In 1862 John Ruskin confided to a correspondent, "I want to believe in Apollo—but can't—the sun is said to be getting rusty (is it not?)."[20] Ruskin may have been reacting to one of thermodynamicist William Thomson's popular articles on solar physics published earlier that year, "On the Age of the Sun's Heat," stating that the authority of physics granted to the sun's prior existence far less time than was posited by Darwin's spacious chronologies of natural evolution. Thomson's best estimate of the age of the sun was somewhere between ten and one hundred million years. "As for the future," he concluded, "we may say, with equal certainty, that inhabitants of the earth cannot continue to enjoy the light and heat essential to their life, for many million years longer, unless sources now unknown to us are prepared in the great storehouse of creation."[21]

Edward Bulwer-Lytton's popular short novel of 1871, *The Coming Race,* is an allegory of energy and evolution that resonates closely with the conservative cultural agenda of the North British thermodynamicists.[22] It imagines a new, previously unknown source for what William Thomson had termed the "light and heat essential to life," suggesting that human life could be possible without sunlight and coalfire, if the (electrical) forces hidden within the physical world can be mastered. The coming race, the Ana, has been exiled for thousands of years within subterranean caverns. But these posthuman beings are no prisoners in a Platonic cave. The most highly evolved groups of the race, the Vril-ya, have achieved superhumanity through consummate technological control of their allotted world. A superior form of energy flows into their "world without a sun" through the channel of *vril.*[23]

Here superhumanity is a state of being brought about by a perfect manipulation of perfected energies. Vril energy is conducted by "a subtle and life-giving medium called Lai," the narrator of the tale tells us, "which I suspect to be identical with the ethereal oxygen of Dr. Lewins, wherein work all the correlative forces united under the name of vril; . . . wherever this medium could be expanded, as it were, sufficiently for the various agencies of vril to have ample play, a temperature congenial to the highest forms of life could be secured."[24] It appears that Lytton modeled vril, at least in a roundabout way, on the same all-pervasive luminiferous or electromagnetic ether through which Maxwell had recently produced a unified field theory of radiant energies. Vril flows through the earth as the ether was conceived to fill all space.

The dedication—"Inscribed to Max Müller in tribute of respect and admiration"—indicates that Lytton's allegory of energy is also an inversion of Victorian solar mythology.[25] Müller's various mythological and linguistic treatises established a reductive hermeneutics of archaic narrative centered on the temporal eclipse of the sun and the similarly entropic drift of metaphor "from *nomina* into *numina*."[26] But whereas Müller's system offers an anthropological allegoresis of myth, *The Coming Race* uses a fictive anthropological report on the race of the Vril-ya to frame an allegorical apology for the politics of empire. Adapting a Müllerian analysis to the language of the Ana, inventing etymological and grammatical structures into which he then embedded his social and political convictions, Lytton roots bourgeois ideology scientistically into the supposed linguistic essence of the superior race.

The Coming Race is occasionally cited as a precursor to Well's *The Time Machine*. Like its more renowned descendent, it too enacts a scientistic response to cultural issues through the mode of allegory. Both stories present fantastic journeys to fantastic underworlds, tableaux in which distinct forms of life occupy different spaces and levels of being. Both are allegories of evolving and devolving social classes mediated by the incursion of an unnamed first-person narrator, a visitor who becomes a cultural go-between. Both deploy evolutionistic motives for the metamorphosis of humanity into daemonic forms: in Wells the spectacle of the variously devolved races of the Eloi and the Morlocks, in Lytton the description of the sublimely tranquil yet imposing and potent Vril-ya. Upon the narrator's first encounter with an An, "it came within a few yards of me, and at the sight and presence of it an indescribable awe and tremor seized me, rooting my feet to the ground":

> It reminded me of symbolical images of Genius or Demon that are seen on Etruscan vases or limned on the walls of Eastern sepulchres—images that borrow the outlines of man, and are yet of another race. It was tall, not gigantic, but tall as the tallest men below the height of giants. Its chief cover-

> ing seemed to me to be composed of large wings folded over its breast and reaching to its knees. . . . It wore on its head a kind of tiara that shone with jewels, and carried in its right hand a slender staff of bright metal like polished steel. But the face! it was that which inspired my awe and my terror. It was the face of man, but yet of a type of man distinct from our known extant races. The nearest approach to it in outline and expression is the face of the sculptured sphinx—so regular in its calm, intellectual, mysterious beauty.[27]

Wells's Time Traveller will also encounter a winged sphinx at the moment of his entry into the era of the Eloi. In Lytton's stark prosopopoeia, the superhuman essence or "type" of the coming race can be read in the archaic lineaments of a monumentally mythic visage. The Vril-ya become more daemonic still when they suddenly spread their wings and launch forth into the luminescent spaces of their subterranean realm. But eventually it is made clear to the narrator that these wings are detachable mechanical prostheses that operate by the control of vril. The Vril-ya have achieved superhumanity by mastering a form of subtle energy and then coevolving with it, integrating their biological and technological being. A full-blown cyborg fantasy leaps directly out of raw mid-nineteenth-century materials—early thermodynamics, social evolutionism, and the concurrent stirrings of cultural response to electromagnetic apparatuses and theories.[28]

There is also in this high Victorian milieu the cultural need to legitimate imperialism through the authority of scientific theories. The fable of the Vril-ya models energy by transferring it from natural to human agency, an analogue of the power technologies driving the transportation and communication systems assisting the transfers of wealth that run the British Empire. In this way, evolutionary status is determined by the technological extent of "human"—scientific and imperial—power to exploit "natural"—aboriginal and colonial—resources. As the narrator learns from the adolescent Taë, who speaks to him using the Vril-ya's terms for "democracy"—*Koom-Posh*—and "universal strife"—*Glek-Nas,* the internally peaceable Vril-ya are not so solicitous about the rights of the indigenous races on the peripheries of their realm:

> Of course, we cannot settle in lands already occupied by the Vril-ya; and if we take the cultivated lands of the other races of Ana, we must utterly destroy the previous inhabitants. Sometimes, as it is, we take waste spots, and find that a troublesome, quarrelsome race of Ana, especially if under the administration of Koom-Posh or Glek-Nas, resents our vicinity, and picks a quarrel with us; then, of course, as menacing our welfare, we destroy it: there is no coming to terms of peace with a race so idiotic that it is always changing the form of government which represents it.[29]

Lytton's fable about a superior underground civilization is first and foremost a plea for imperial fortitude and a cautionary statement about the domestic and overseas perils of "equilibration," that is, liberal democratization. The narrator is a proud but benighted Yankee democrat who receives from the politically perfected Vril-ya, much like Gulliver from the Houyhnhnms, doctrinal deprogramming. Scientific ideas enter the text precisely as ideological counters for this sociopolitical message.[30] Lytton uses a series of evolutionistic bourgeois scientisms to construct a racist allegory about political authority—covered over in the fable as the biological capacity and spiritual power to master energy. As one might expect from the title, in Lytton's fable the authority to control vril is determined by race— "hereditarily transmitted organisation."[31] Among the Ana, over many thousands of years the Vril-ya have so perfected vril technology that they have evolved a new nerve to interface with the "Vril Staff." After describing the powers exercised through the Vril Staff by Zee—an erudite and accomplished young Gy, or woman—the narrator focuses on the hand of this phallic maiden:

> In the first place, the thumb of the Gy (and, as I afterwards noticed, of all that race, male or female) was much larger, at once longer and more massive, than is found with our species above ground. . . . More remarkable . . . is a visible nerve, perceptible under the skin, which starts from the wrist skirting the ball of the thumb, and branching, fork-like, at the roots of the fore and middle fingers. . . . "That is not found in the hands of our earliest progenitors, nor in those of the ruder tribes without the pale of the Vril-ya. It has been slowly developed in the course of generations, commencing in the early achievements, and increasing with the continuous exercise, of the vril power."[32]

In sum, vril is an allegorical energy form through which *The Coming Race* can channel the prestige of currently authoritative physical concepts toward particular political ideas. Vril itself—an infinite energy conserved throughout its willed transformations—is figured by a recuperative reading of the first law of thermodynamics. Power over vril is thus the truly *conservative* state of being. Seizing this energy the Vril-ya become the gods of their own substitute sun: pure, constant, imperious priests of vril. The social affect of the second law of thermodynamics is then displaced upon the signs of political liberalism, Koom-posh and Glek-nas, democracy and "the universal strife-rot."[33] Upon democracy, especially in its most "highly evolved" American incarnation, Lytton displaces the anxieties traditionally associated with physical and physiological ideas of disorder, corruption, and chaos, and recently amplified by the correlated scientisms of entropy and degeneration. But for good measure, Lytton adds another evolutionistic twist to these allegories of energy. Against the dissipation of capital and the chaotic redistribution of authority pursued under liberal democratic regimes, the novel offers a Spencerian or social Darwinist reading of natural selection as the fatal mainspring of the superracial aristocracy of the Vril-ya. In addition, their prophesied return to and reconquest of the upper world melds Platonic myth to imperial motives:

> The superiority of the Vril-ya was supposed to have originated in the intensity of their earlier struggles against obstacles in nature amidst the localities in which they had first settled. "Wherever," said Zee, moralising, "wherever goes on that early process in the history of civilization, by which life is made a struggle, in which the individual has to put forth all his powers to compete with his fellow, we invariably find this result—viz., since in the competition a vast number must perish, nature selects for preservation only the strongest specimens. With our race, therefore, even before the discovery of vril, only the highest organizations were preserved; and there is among our ancient books a legend, once popularly believed, that we were driven from a region that seems to denote the world you come from, in order to perfect our condition and attain to the purest elimination of our species by the severity of the struggles our forefathers underwent; and that, when our education shall become finally completed, we are destined to return to the upper world, and supplant all the inferior races now existing therein." (62)

This is why the Vril-ya are the *coming* race: they have purchased their perfection through the political wisdom proper to an island existence, a tribal, patriarchal isolation. They will find the dissipated races of liberal Europe and America easy to eliminate when they finally emerge to colonize the sunlit world and "utterly destroy the previous inhabitants." This vision of the future ruin of the present human chaos is the novel's final metonymy of thermodynamic entropy.

Phantasmagoric Speculations: *La Fin du Monde*

> For at least 200 years, say from the time of Newton to the late nineteenth century, the dominant preoccupation of science was with those chains of cause and effect which could be referred to forces and impacts.
>
> —Bateson, *Steps to an Ecology of Mind*

Ominous bodily fatigue and predictions of cosmic waste were the cultural underside of the superhuman effectivity of modern energies and their cascading technological applications. The frictions of industrial exhaustion were themselves driven by the new technologies of physical power. Throughout the nineteenth century, the cultural image as well as the actual workings of material forces heated up, became ever more grandiose and explosive. The experimental verification of the speed of light, as well as the impressive mundane velocities of artillery shells and locomotives, provided increasing incentives to envision and calculate the forces of masses in rapid motion through immense spaces. During this period, from the molecular to the galactic level, in some effective or spectacular fashion, the speeding masses of the late Newtonian cosmos are increasingly envisioned in spectacular collision.

In Flammarion's *La Fin du Monde* of 1893, the specter of a cosmic collision between the earth and a killer comet offers emblematical cover for a number of urgent sociopolitical contents. In this section I will trace the way that this fin de siècle apocalypse uses scientific discourse to provide phantasmagoric mediation for a series of worldly events—war, revolution, and the economy. To begin with, *La Fin du Monde* is fairly explicit about the martial connotations of the impending cosmic threat. At first Flammarion's comet appears like "an army of flaming meteors"[34] (fig. 9). But it is sufficiently gaseous that "the main body of the celestial army" (177) passes by without causing a total devastation. Moreover, despite some momentary damage, the comet's ultimate effect is beneficial, reinvigorating: "The losses were made good by an apparent increase in human vitality, such as had been observed formerly after destructive wars; the earth continued to revolve in the light of the sun, and humanity to advance toward a still higher destiny" (185). Flammarion's similes underscore *La Fin du Monde*'s ideological drive, despite its manifest critique of standing armies, to justify capitalist war economies by positing its dialectical necessity as a negative moment leading to a future evolutionary transcendence of war altogether. The losses of combat and the expenditures demanded by war economies are recuperated and rationalized through a wishful reading of dynamical regimes in which sudden bouts of destruction yield a creative restructuring at a higher level of development.

Fables of impending destruction by colliding planets can also mask the direct collision and fiery clash of political regimes in a social revolution. Flammarion develops the world of the twenty-fifth century as the progressive aftermath of a series of revolutionary transformations, beginning with "the great social revolution of the international anarchists who, in 1950, had blown up the greater portion of [Paris] as from the vent of a crater."[35] Flammarion's own political attitude is discernible as a fairly positivist, scientized form of libertarianism. To banish the military and purge the state bureaucracy once and for all, Flammarion tells us, "a radical revolution was necessary. From that time on, Europe had advanced as by magic in a marvellous progress—social, scientific, artistic, and industrial. Taxation, dimin-

Fig. 9. Une pluie de bolides s'abattit. (A rain of projectiles beats down.) From *La Fin du Monde.*

ished by nine-tenths, served only for the maintenance of internal order, the security of life and property, the support of schools, and the encouragement of new researches. But individual initiative was far more effective than the old-time official centralization" (194–95). The passage of the comet is the last in a series of counter/revolutionary preparations and the final catalyst for humanity's higher evolutionary development.

At the beginning of the narrative, however, Flammarion also treats the emergence of the comet in its aspect as news, that is, as a doomsday media item hyped by the popular press, and as a modern telecommunicated disaster, reported more or less in real time. The charge of the extraordinary colloquium at the Parisian Institute is to separate vulgar misunderstanding and media misinformation from a properly scientific grasp of the situation. Like *Wall Street Week* pundits predicting the coming year's movement of the Nasdaq average, the Parisian experts engage in studied debate about the impending comet's potential effects. The first speaker reassures the audience that the world will not be utterly destroyed by the comet and illustrates his confidence with the following declaration: "I confess that if the bourse was not closed, and if I had never had the misfortune to be interested in speculation, I should not hesitate today to purchase securities which have fallen so low."[36] However, he is abruptly interrupted in a tendentious fashion: "This sentence was not finished before a noted American Israelite—a prince of finance—director of the journal *The Twenty-fifth Century*, occupying a seat on one of the upper steps of the amphitheater, forced his way, one hardly knows how, through the rows of benches, and rolled like a ball to the corridor leading to an exit, through which he disappeared."[37] Flammarion's illustrator subtly depicts this "prince of finance" as a Prince of Darkness (fig. 10). He rolls out of the domed meeting hall, the tails of his frock coat spreading out to suggest the wings of a fallen angel embarking on a devilish scheme. Dashing down like a diabolical fireball, this Luciferian manipulator is a refraction of the evil comet looming overhead. He wings away by airship to Chicago and places his new scientific intelligence into timely motion. Capitalizing on the information that the comet crash will not be globally fatal, he causes the stock markets to be reopened and buys up shares whose price has crashed in mistaken anticipation of the cosmic catastrophe.

Flammarion's text provides a skewed but acute analysis of media manipulation, the ways that the public traffic in scientific assertions can impinge on a market economy. "As in other days," Flammarion informs us, "the monasteries had accepted bequests made in view of the end of the world, so our indefatigable speculator had thought best to remain at his telephone, which he had caused to be taken down for the nonce into a vast subterranean gallery, hermetically closed" with "special wires uniting . . . the principal cities of the world."[38] From this occulted command-center the devilish market-and-media manipulator makes a further fortune off the comet affair. At first, responding to an unverified local report, headlines in *The Twenty-fifth Century* declare, "Death of the Pope and all the bishops! Fall of the Comet at Rome!" (177). When it turns out that the Vatican has been unharmed, the wily editor provokes further circulation by cynically con-

Fig. 10. Se précipita comme une boule. (He rolls down like a ball.) From *La Fin du Monde*.

verting the correction of misinformation into an act of divine intervention: "Great Miracle at Rome!" (181).

In the bourgeois allegory at stake in these passages, the crucial connections are those among energy, capital, and catastrophe. For the ominous comet looms over this story like the threat of a stock market collapse over a mismanaged or overheated economy. Indeed, capital investment is that to which (financial) catastrophe can occur—and in the form precisely of a *crash,* the collision of high prices falling back to earth, as "ruined" speculators jump to their deaths out of high windows. And just as some thermodynamic moralists bank on the "impending heat death of the universe"—that is, on some train of reasoning based on the *imminent* possibility of some universal entropic catastrophe—one can capitalize *on* a stock market crash, especially on the mediated perception that such a crash is "impending," due at any moment. And yet, the market generally "revives" after a crash; as in these cosmic collision scenarios, the momentary devastation clears the ground for a utopian future.

Within *La Fin du Monde* the stock market affair is a bit of social satire leavening the heaviness of the narrative's other apocalyptic themes. Flammarion demonizes the Jewish tycoon with a scenario of cynical exploitation. "Speculation" is explicitly presented as the corrupt capitalistic device of playing on manufactured perceptions, especially on expectations of the future. In the comet's aftermath, with the dissolution of the military in the onset of a more perfect society, Flammarion explains that "men perceived that the military system meant the maintenance of an army of parasites and idlers, yielding a passive obedience to the orders of diplomats, who were simply *speculating* on human credulity."[39] Through the Jewish tycoon and the international diplomat, Flammarion defines and demonizes a culture of destructive speculation, developed to paranoid length by details such as the subterranean stronghold in telecommunication with minions located in military-industrial capitals. But this satire actually bites the author back by revealing—through the projection of his daemonic double, the "American Israelite"—the structure of his own rhetorical manipulations.

The ideological alarm associating Jewish financiers with market capitalism and media corruption centers on the term *speculation* in a way that incriminates Flammarion as well. For the entire scenario of financial speculation emerges in tandem with the comet's impending and actual collision with the earth, which main event is itself produced by an act of *fictive* speculation, the very "futuristic vision" that holds this narrative in the grip of thermodynamic images of catastrophe coined earlier in the nineteenth century. In the phantasmagorias of thermodynamic apocalypse that emerged in the 1850s and 1860s, entropy came forward as the uncontrollable aspect of physical systems, historically concurrent with other modern anxieties over the control of capital, material resources, ethnic minorities, and subject populations. The initial notion of thermodynamic entropy was charged at its inception by the cultural stresses of imperialism and industrial capitalism, and then elaborated by later narrations that reproduced that socioeconomic stress under the name of physical entropy, providing a paranoid view of present social relations extrapolated into a speculative future.

Flammarion articulated some of the forms of mythic history and cultural paranoia induced by the initial discourse of thermodynamics—the notion that somehow the *cosmos* (as opposed to the economy as operated by a particular political regime) is out to get us, perhaps by dooming us to the chronic fatigue syndrome of a slow entropic death, or else by hurling at us out of nowhere a fiery comet of vast destruction. In these equivocations on the act of *speculation,* financial, scientific, and fictional predictions typically produce an allegory of the present, a transformation that presents the present under threat of daemonic control by the future. The paradigmatic science-fiction of a speculative future necessarily re-presents the chaos of the present in the mode of allegory—as a totalized system—and this textual conversion of the present is then registered within the text of the future as a paranoia of control. In *La Fin du Monde,* allegory and commodification converge to promote Flammarion's own brand of prophetic positivism, to enhance his own speculative holdings of intellectual paper, through phantasmagoric spins on the laws of thermodynamics.

In our culture, the discursive products of science inevitably enter into allegorical dialects of commodification. Within the "axial fields" of Benjamin's dialectical coordinates, the cultural allegories instigated by the natural sciences stake out various mythic and historical positions. The discourse of evolution developed into an allegory of industrial society, as the fossil or trace of natural history was transferred to the obsolescence of used goods and outmoded ideas. Commodity fetishism marks the phantasmagorias that unite scientific ideas to social demands, that translate material into mythic history. Wish images circulate as symbols of a mythic nature transfused in modernity by the awesome new divinities of the heat engine and the electrical dynamo. In the era of classical thermodynamics, the emblematical ruin of historical nature returns in the iconography of entropy, heat death, and crashing comets. Yet these early phantasmagorias of energic or entropic catastrophe show that scientistic prophecies of disaster are always local rhetorical manipulations. One advocates the bitter doom that most sweetens one's ongoing situation.

Chapter 3
Technoscientism

Technoscientism as Cultural Allegory

Scientism has formerly been conceptualized according to what Bruno Latour calls the *diffusion* model of the sociological field. In this model, science is privileged over other discourses as a ground of epistemological origin and sealed off from the irrationality of the rest of culture. On the diffusion model, scientism is an epiphenomenon of real science, the relatively irrational social excess of scientific production. The diffusion model enforces a regime of sociological separatism; it produces an erroneous "*belief* in the existence of a society separated from technoscience."[1] This diffusionist model of scientism is itself a relic of the social sciences' own scientistic bid for intellectual authority. Iain Cameron and David O. Edge's classic exposition of scientism operates within the diffusionist framework, in which all scientistic commerce is conceived as a one-way outflow from a privileged scientific source: "Scientism is present where people draw on widely shared images and notions about the scientific community and its beliefs and practices in order to add weight to arguments they are advancing, or to practices they are promoting. . . . Those who use scientistic language acknowledge and respect *the authority of the scientific community*, and wish to capitalize on that authority. . . . In so doing, they reinforce and consolidate that authority."[2]

From this perspective, the modern proliferation of physical, biological, and mathematical scientisms is seen as a relatively vitiated part of the cultural interaction among well-demarcated disciplinary realms of science, technology, and society. Diffusionist scientisms are illegitimate offspring of science that cobble to a extrascientific object the cultural aura of science's own epistemological prestige, typically by an extension of scientific terminology, imagery, and/or methodology. This adaptation can be deliberate or unwitting, earnest or satirical. It can be more or less successful as an intellectual or a social gambit. But irrespective of its discursive success or failure, scientism on the diffusion model is a sloppy affair, the random "spillover" of science into society, an inappropriate and at worst abusive extension of scientific terms or practices.

For instance, the pioneering sociologist Max Weber "criticized positivism and scientific naturalism, singling out the social energeticists—Ernest Solvay and Wilhelm Ostwald—for their 'umstülpung,' or spillover, of the 'world picture' of scientific disciplines into the 'worldviews' of the social sciences, where they ought not have a place."[3] On the diffusion model, such extensions of scientific concepts are illicit displacements of authority from science to nonscientific matters, a transfer, sometimes a plunder, of science's social prestige. Scientisms are aberrant discourses circulating through social channels free of scientific control yet demanding some level of literal credence for what are at best overworked figurative conceptions. These abuses bottom out in "pseudoscience"—bogus representations or active misuses of scientific ideas. In sum, as a repository for derivative and deformed conceptions *scientism* has typically been a term of bad intellectual repute.

However, no cordon sanitaire actually keeps technoscience apart from its social causes and effects. As Donna Haraway puts it, "Technoscience is civics, in the strong sense, at the heart of what can count as knowledge."[4] A diffusionist conception of scientism is no longer tenable. In fact, most scientism is performed in good faith, even if it has bad results. Much of it subsists as obsolete or premature scientific protocols deriving from proper insights that arrived in inappropriate locations. For instance, Gregory Bateson saw the vogue for physical energy models in the behavioral sciences as the problematic extension of a kind of intrascientific scientism.[5] Hoping to update and replace such mechanical energy models, for example, neural "impulses," with the cybernetic idea of informatic ensembles, what he called "transforms of differences," Bateson's polemic was aimed at the problematic effects of physical scientisms not just in the culture at large, but precisely in those models of psychological and psychiatric science propped upon classical thermodynamical paradigms. Merely derogatory or separatistic definitions of scientism, therefore, often obscure the complicity of technoscientism with scientific productions.

As Michel Serres has often argued, the most prevalent and injurious myth about science is the idea of a science without myth. By its own heuristic methods technoscience participates in the formation and dissemination of scientisms that drive both scientific and cultural developments. Technoscientisms include the positivist variety, as with the vigorous extension of physical energy models into social and psychological domains. But these dominant positivist scientisms are also joined by all manner of creative reworkings of scientific concepts in various aesthetic and discursive genres, and these visionary and speculative scientisms, I will argue throughout this study, may possess critical, rhetorical, or practical merits irrespective of their epistemological status.

The era of classical thermodynamics was enthusiastic and far-reaching in its adaptations of science to other realms of cultural production. Surveying the role of creative scientisms in the rise of early-modernist abstraction in the visual arts, Luc Ferry writes: "In an atmosphere heavily impregnated with scientism, . . . painters had the feeling that a new scientific age legitimized a new artistic age."[6] For another seminal instance of positivist scientism in coordinated action with sociopolitical critique, Karl Marx's preface to *Capital* develops a series of biological, physical, and chemical tropes in support of its political and economic arguments. Responding to the mid-nineteenth-century intellectual glamor of cellular and evolutionary biology, Marx illustrated the social function of commodities through biological metaphors. Commodities represent the "cells" of the socioeconomic "body": "in bourgeois society the commodity-form of the product of labor—or the value-form of the commodity—is the economic cell-form."[7] This microeconomic scientism is then coordinated with the progressive promise of evolutionary theory by reading the macroeconomy as an "evolutionary" population.

Marx explains that in his system, "the evolution of the economic function of society is viewed as a process of natural history."[8] As a result, just as in the larger sweep of "natural history"—that is, evolutionary development—the vagaries of biological individuals are of statistical insignificance, so "here individuals are dealt with only in so far as they are the personifications of economic categories, embodiments of particular class relations and class interests" (417). Marx asserts the scientificity of his project precisely by flattening individual persons into corporate personifications. As Walter Benjamin would have noted, his appeal to science is at the same time an appeal to allegory, that durable resource of emblematic figuration wherein "human figures" are the impersonal signifiers of (economic or scientific) concepts.

My point is that the scientism of Marx's discursive system neither validates nor invalidates it: rather, it historicizes and so relativizes it. Put another way, scientisms produce a nesting of discrete systems whose overlap will work in some ways but not in others. To pursue a less prejudicial and accusatory historiography of scientism as a rhetorical and discursive force, we can refer to Bruno Latour's *translation* model of socioscientific feedback: "In the translation model, there is *no transportation without transformation*."[9] In other words, critical explanation must begin from the elements and processes of mediation within the total cultural field—in our dialect, the modal circuits constructed by the many *allegories* of science. The translation model would render scientism as *techno*scientism, an epistemologically ambivalent and historically dynamic body of heuristic vehicles and visionary hunches embedded in heterogeneous yet interconnected cultural matrices. Technoscientisms enroll adherents and construct alliances through mythopoetic resonance as well as scientific aura. A discourse on technoscientism would operate in the manner of Michel Callon's "principle of generalized symmetry," in which "the anthropologist has to position himself at the median point where he can follow the attribution of both nonhuman and human properties"; as Latour puts it succinctly, under this epistemological regime, "the explanation starts from quasi-objects."[10] The generalized symmetry of technoscientism, then, prevents the facile attribution of truth or falsehood to any construction of nature or society, seeing all such productions as hybrid assemblages operating within stronger or weaker networks or collectives.

We will say that technoscientism in its various guises is a mobile and pervasive *allegory of science.* Allegory has in fact been the perennial modal marker of scientistic discourses, the prime narrative and pictorial medium through which scientisms operate as tokens of power. Reification, personification, typological allusion, and emblematic juxtaposition form a significant part of the arsenal of tropes and schemes that both shape up and ship out scientific ideas, imparting semantic content, figural structure, narrative momentum, and at times, satiric and critical contingencies, to the stories the sciences tell. As we have been tracing through chapters 1 and 2, the standard vehicles by which the sciences are conveyed into public and popular discourse, by which the sciences are skewed, even in their inception, toward private and parochial interests, are modes of verbal or visual allegory.[11] Allegory couples with science to spawn a motley but culturally legitimate offspring of technoscientistic figurations.

Exploring a wide range of scientistic productions deriving immediately from contemporary technoscience, Donna Haraway has indicated their allegorical dimension through Erich Auerbach's term *figuration:* "I consider figures to be potent, embodied—incarnated, if you will—fictions that collect up the people in a story that tends to fulfillment, to an ending that redeems and restores meaning in a salvation history. . . . Contemporary technoscience in the United States is deeply engaged in producing such stories, slightly modified to fit the conventions of secular realism."[12] Other examples of contemporary technoscientism at work are the cultural field models prominent in recent sociological and literary discussion. For instance, Martin Jay's essay "Modernism and the Specter of Psychologism" presents a socioliterary field model of intersecting discursive planes: "the complex and heterogeneous cultural field we have come to call aesthetic modernism" possesses "multiple intersections with that other richly articulated field known as modern psychology."[13] Jay goes on to discuss a series of modernist responses and resistances to the extension of psychologistic models to other areas of cultural production. However, his own adaptation and discursive extension of the physical field model to the narration of cultural history is a standard kind of general scientism applied to critical discussion.

Jay's heuristic evocation of the cultural field, however, pales before the extensive and reified development of a physical field model in Pierre Bourdieu's *The Field of Cultural Production:*

> The science of the literary field is a form of *analysis situs* which establishes that each position . . . is subjectively defined by the system of distinctive properties by which it can be situated relative to other positions; that every position, even the dominant one, depends for its very existence, and for the determinations it imposes on its occupant, on the other positions constituting the field; and that the structure of the field, i.e. of the space of positions, is nothing other than the structure of the distribution of capital of specific properties which governs success in the field and the winning of the external or specific profits (such as literary prestige) which are at stake in the field. . . . The literary or artistic field is a *field of forces,* but it is also a *field of struggles* tending to transform or conserve this field of forces.[14]

On this basis Bourdieu constructs a fluid sociological schema of nested systems, in which the literary field and the artistic field float within a larger formation, the field of power, which is itself a mobile assemblage within the field of class relations. Bourdieu emphasizes the charged relations that develop between a dominant "heteronomy" and a dominated "autonomy," within and among all the fields of the social matrix. For instance, in the literary and artistic fields the heteronomic pole is bourgeois or profitable art, the autonomic pole avant-garde or nonprofitable art. The social dynamism of "symbolic capital" charges the cultural field with (symbolic) energy, and accounts for the convoluted inversions that mark artistic fields, as when the dominated (autonomous) artist lines up at the negative pole of the field of power but the positive (prestigious) pole of the field of cultural producers.

In general, as cultural extensions of allegorical devices, technoscientisms are woven from discursive cross-purposes—celebration and criticism, tradition and innovation, demonstration and speculation, and so forth. Like Jay, Bourdieu critiques one form of scientism while performing another. By factoring the symbolic capital of social prestige into the total circulation of cultural "energies," the polar inversions within the matrix of cultural fields undermine strictly economistic models. Nevertheless, his analytical model is itself a technoscientism of considerable proportions—that is, a critical model of "social capital" as cultural energy. This equivocal status does not necessarily diminish the utility of Bourdieu's field models for the analysis of the social capital of scientific or other ideas. At the same time, Bourdieu's field model continues the allegory of energy as capital and commodity we examined in the previous chapter.

In the preface of *Chaos Bound,* Katherine Hayles remarks that in the physical concept of an energy field, "reality consists not of discrete objects located in space but rather of an underlying field whose interactions *produce* both objects and space."[15] In a physical field, an ensemble of forces—electromagnetic or gravitational energies—constitutes and so interrelates the objects distributed within it. Hayles then concretizes the physical field concept in its technological manifestation as a metaphor for the discursive transformations of the current episteme under the regimen of modern communications. "Complex connections between literature and science," Hayles writes, "are mediated through the cultural matrix. . . . Cultural fields bespeak the interconnectedness of a world in which instantaneous global communications is a mundane reality" (4). So in the cultural field, discrete disciplines are investigated within the informatic network of their discursive interdependencies. With Hayles, however, the critical scientisms involved in the extension of the field and dynamical-systems models are held constantly in mind and interrogated as they are deployed. The cultural field metaphor allows Hayles to construct the relations of literature and science in a way that withdraws the priority granted to science in previous academic scientisms. In this schema, all disciplinary sites must negotiate their relations to other sites through an ongoing tension of transfers and resistances. Hayles's field models gain critical power to the extent that they operate in full cognizance of their allegorical status as heuristic rhetorical operators.

Technoscientism, then, is a monumental allegory of modern science produced by the transdisciplinary circulation of scientific metaphors and, in field theories in particular, by the formal extension of models of energy. The very movement of concepts from site to site within the cultural field tends to curve signs into polysemous figurations, multileveled allegories. Even within a single disciplinary site, the use of heuristic figures can bring about some productive increment of discursive excess. Hayles discusses this phenomenon with regard to the heuristic career of Maxwell's Demon: "the detour through language which the heuristic represents"—in this instance, the placing of the overdetermined term *demon* into this physical context—"creates polysemous connections not present in the theory itself."[16] The kinetic theory of gases and the second law of thermodynamics, mundane phenomena of heat flow, become literally daemonic—that is, creative of new discursive mediations—because "overlaying a heuristic onto a theory is never merely an inert transposition of concepts . . . it generates a surplus of signification" (33).

Hayles nicely states here what I take to be the abstract rhetorical basis of technoscientism. The scientistic allegorization of science arises from the excess of epistemic signification inherent in any social semanticizing of scientific or mathematical ideas. Generated within given disciplines, the excess signification of heuristic theorizations elicits speculative extensions, in William Paulson's trenchant phrase, a productive *cultural noise*—in various critical and creative forms—within the channels of technoscientific communication. Scientism is a dynamic and unpredictable irruption within the noise of technoscience, giving rise sometimes to dubious social or political doctrines, sometimes to striking productions of artistic and critical self-organization, sometimes to reorganizations of the entire cultural field. Scientism is the prolific and systematically inevitable supplement of technoscientific action.

Thermodynamic Scientisms

In Britain the phenomena that later came to be associated with the term *entropy* were first gathered under the names of *dissipation* and *equilibration*. Hayles has commented on the powerful crossover between physical and moral registers inscribed in the equivocal term *dissipation*—its function, in our dialect, as a scientistic operator: "The 'universal tendency toward dissipation' places entropic heat loss in the same semantic category as deplorable personal habits."[17] At the same time, the semantic ambiguity of "dissipation" encouraged Victorian thinkers to contemplate the possibility that, like a dissolute individual, the race or the cosmos could be reformed, rehabilitated, or in some fashion redeemed from its wasteful ways, by some development or change in human knowledge or being. A broad cultural imperative in response to thermodynamic moralism was thus an ethical demand for some form of spiritual reordering, some free-willed motion of antidissipative fortitude. Still strongly tied to the tradition of natural theology, the earliest discourses of Victorian thermodynamics were moralized through a litany of Christian injunctions.[18]

In his seminal paper of 1852, William Thomson formulated the universal outcome of the second law of thermodynamics as the "dissipation of mechanical energy": as a result, "within a finite period of time to come the earth must [be] unfit for the habitation of man as at present constituted."[19] Unless God defied the first law and intervened with a new creation, the not-so-distant future appeared to hold only an eternity of dead matter wheeling meaninglessly through space at three degrees above absolute zero, an apocalyptic prediction of cosmological consequences, breathtakingly extended from an unassuming physical postulate about the thermal motions of molecules—that unless external agency is applied, heat always flows from warmer to cooler bodies. The model of energy presented by this vision of universal heat death is perhaps the supreme expression of the Victorian allegory of thermodynamics. The North British scientists of energy fashioned an influential discourse of thermodynamic scientism that infused physical concepts—energy, dissipation, and equilibration—with moral contents—life, sin, and death.[20]

Early moralizations of thermodynamics were also part of the nineteenth-century vogue for global continuities—a conceptual conviction that greatly aided the formation of vigorous scientisms. The ontological gaps between physical and moral laws, material and cultural effects, were either dismissed or finessed through crude or refined forms of dialectical resolution.[21] Smith notes that for Maxwell and Thomson, "dissipation of energy . . . linked together the natural and moral orders."[22] In 1868 their colleagues Balfour Stewart and Norman Lockyer penned an article, "The Place of Life in a Universe of Energy," that offered a similar dose of energy theology, collapsing the physical and the social and placing cosmic as well as personal redemption in the hands of God: "As in the social world a man may degrade his energy, so in the physical world energy may be degraded; in both worlds, when degradation is once accomplished, a complete recovery would appear to be impossible, unless energy of a superior form be communicated from without."[23]

The thermodynamic dispensations of such theological scientisms were explicitly formulated to brand secular and progressivist social doctrines, in particular, evolutionary scientisms, as unscientific. Yet although the science of energy developed in explicit opposition to doctrines of "scientific materialism" such Herbert Spencer's evolutionist sociology,[24] both discourses exhibited comparable equivocations of nature and culture. For X-Club members such as John Tyndall and Thomas Huxley, Spencer's *First Principles* offered a positivist alternative to the natural-theological schemas of Maxwell, Thomson, and Tait. However, Spencer's intellectual commitment to an immediate continuity between the physical and social spheres produced some severe cosmological dilemmas. His cultural double-bind between "evolution" and "dissipation" is dramatically evident in a letter to

Tyndall in 1858. Spencer was responding to Tyndall's explanation that the rule of mechanical dissipation of energy would lead to an "equilibration" that coincided with universal heat death:

> In the first place I fully recognize, and have all along recognized, the tendency to ultimate equilibrium. . . . The universal tendency towards equilibrium must apply to human actions as well as to others. All our hourly doings are endeavours to equilibrate our acts and wants. Those modifications of faculty which we are ever undergoing are a continuous approximation to equilibrium between constitution and conditions. . . . There must be change until equilibration is reached. This applies to man in his social relations. Society must go on changing until all men's constitutions and their circumstances are in correspondence—*until the desires of each are such only as may be gratified without thwarting the desire of others*. . . . Thus you see that my views commit me most fully to the doctrine of ultimate equilibration.
>
> That which was new to me in your position enunciated last June, and again on Saturday, was that equilibration was death. Regarding as I had done, equilibration as the ultimate and *highest* state of society, I had assumed it to be not only the ultimate but also the highest state of the universe. And your assertion that when equilibrium was reached life must cease, staggered me. Indeed not seeing my way out of the conclusion, I remember being out of spirits for some days afterwards. I still feel unsettled about the matter and should like some day to discuss it with you.[25]

For William Thomson and his circle, because the natural world was accepted as fallen and fated to decay, the second law came as a physical confirmation of a moral truth, a veritable illustration of the 102nd Psalm.[26] The physical and metaphysical ambiguities of "dissipation" could thus be mapped onto the theological dualism of matter and spirit. But for the secular humanist Spencer, civic institutions were fallen, and social evolutionism offered hope that they could be ameliorated by closer coordination with the progressive unfolding of natural processes. In this instance, however, the very systematic equivocations he used to eliminate the gap between nature and culture now tied his vision of evolving culture to the specter of a dying nature. Previously positing "equilibration" as the telos of his evolutionary scheme, Spencer had allowed the concept to circulate at large from individual, to social, to physical and biological development. The notion of energy dissipation leading to thermodynamic equilibrium, even as mediated by the relatively secular Tyndall, now came to Spencer as an unwelcome admonishment, placing a chill on his rosy evolutionary scenario.

Anson Rabinbach's *The Human Motor* provides a helpful breakdown of the interconnected modalities of thermodynamic scientisms as they proliferated in the decades following Thomson's and Spencer's generation. In its epistemological implications, late-classical thermodynamicism subscribed to "positivism—the cognitive monopoly and idealization of the scientific method along with the search for general laws of both nature and society."[27] In its broadest cosmological formation as a theory of universal nature, thermodynamics gave rise to "modern *productivism*—the belief that human society and nature are linked by the primacy and identity of all productive activity. . . . The cosmos was essentially a system of production whose product was the universal *Kraft* necessary to power the engines of nature and society" (3). Thermodynamic scientisms also proliferated within social and political discourses: "The dynamic language of energy was . . . central to many utopian social and political ideologies of the early twentieth century: Taylorism, bolshevism, and fascism" (2). And in its physiological adaptation, classical thermodynamics reconceived the organic body as a machine, a motor or heat engine: "For physiologists armed with the principles of thermodynamics, the energy of the body was not merely analogous to other natural physical forces, it became one among them" (46).

In this extended biological register, physical scientisms based on thermodynamics ran into some pronounced social frictions: "accompanying the discovery of energy conservation and entropy was the endemic disorder of fatigue—the most evident and persistent reminder of the body's intractable resistance to unlimited progress and productivity."[28] The cultural repercussions of classical thermodynamics on the moralization of biological functions linger to this day. For instance, in a 1995 episode of Matt Groening's comic strip *Life in Hell,* Akbar and Jeff each try to stick the other with a tiresome household task. Both claim to be the more exhausted, and their escalating agon of fatigue spirals from the level of the individual—"I've been tired my whole life"—to that of human society—"I'm tired because I carry the world's sorrows on my shoulders"—to the status of the universe. When Akbar laments, "I'm tired because I feel a cosmic connection to the concept of entropy," Jeff one-ups him by unfolding out of the discourse of thermodynamic entropy its promised end: "I'm tired because on a deeply personal level, I can relate to the impending heat death of the universe." The final specter of this imminent and ineffable catastrophe shuts the argument down by sanctioning a certain microcosm of shared inaction: "Fine. Neither of us will change the catbox."

Groening's deadpan spoof of the cosmic lengths we sometimes go to avoid unpleasant work is also a cultural time-capsule, preserving in the amber of late-twentieth-century ephemeral humor the anachronistic yet still-viable discursive artifacts of classical thermodynamics. During this era the psychology of stress was referred to the physiology of fatigue, which in turn was modeled on the dynamics of entropic physical systems. With these relays in place, a heavy-industrial civilization invested scientific cachet in the projection of physical laws upon the socioeconomic labor of human beings.[29] The particular thermodynamic discourse of entropy and heat death Groening retails in *Life in Hell* has in fact been marked for over a century as demonic or "hellish" especially in its "cosmic connection," the scientistic vision that pairs physiological to cosmological exhaustion, once all heat differentials die away, to damn the material universe to an ultimate and eternal energic collapse.

Energetics and Vitalism

C. C. Gillispie concluded his history of modern science, *The Edge of Objectivity,* with an attack on what he perceived as an abusive turn-of-the-century strain of thermodynamic scientism: "If the robber baron and the strong man armed took license to ruthlessness from the theory of natural selection, the intellectual who shrank from such successes found in entropy the excuse to indulge his *mal du siècle.*"[30] Writing in the late 1950s, Gillispie focused briefly on American historian Henry Adams to economize his condemnation of the effete despair of a certain strain of modernist intellectual culture: "To Americans, the voice of Henry Adams comes as the most familiar of those speaking out disillusionment from the soured dream of progress. Instead of moving toward the perfection of the race, history runs downhill toward a miscellany of the second-rate, a society without classes, structures, gradations, duties, responsibilities, nobility, or culture, without Mont-Saint-Michel or Chartres" (405).

Gillispie's disdain of this morose school of social entropy provides a rhetorical flourish for the conclusion of his positivist history of scientific progress. He thus distanced his privileged topic from some bothersome scientistic repercussions of that history, various erroneous appeals to "nature" that attacked more liberal visions of social democracy. In Gillispie's version of Adams's ideas, "The future holds only equality and social randomness. But surely Adams had despaired of liberalism, and of his education, before he lost faith in the universe? Surely he, and his literary contemporaries accepted with something very like irresponsibility and relief the imminent prospect of an end to time, when though the masses would be one with the fastidious, none would perceive the universal degradation."[31] Gillispie figured Adams and other litterateurs as a "fastidious" elite whose social punditry amounted to a perversion of scientific enlightenment.

But in each case—the liberal association of intellectual "decay" with a vitiated upper class, the conservative vision of moral "dissipation" amid the lower classes, the socialist equation of cultural "degeneration" with an unrestrained bourgeoisie—the political argument turns on a larger allegory of thermodynamics running through modern discourse. Each is a metaphoric projection of class dynamics upon the thermodynamic differentials between "high" and "low" energies, a political moralization of the physical economies of conservation and waste. Gillispie's liberal positivism may well be preferable to Adams's *mal de siècle,* but his "classifying" of intellectual work is no less ideological a stance.

The contradiction masked by Gillispie's own ideology is that within the text of the culture he inhabits in common with Adams, the sciences are inextricably tangled together with their scientisms. Despite Gillispie's attempt to separate them, the sciences cannot be so absolved. The scientific expositors of thermodynamics have not been innocent of the social shapes thermodynamic ideas have taken. In Henry Adams's most detailed discussion of thermodynamics as a cultural problem, *A Letter to American Teachers of History*—a lengthy freelance report to the American Historical Association—Adams reviewed turn-of-the-century science, first of all, in order to pose thoughtful questions about university administration and the hierarchy of knowledge, as well as to quail before the specter of heat death. Ostensibly at least, he codified his physical anxieties not to evade but to take responsibility for the transmission of science into the discourse of the humanities. Assembling the testimonies of a series of prominent physical, biological, sociological, and cosmological authors, Adams's *Letter* brings the diffuse socioscientific malaise of his time into clear focus.

At the turn of the twentieth century, the cultural vogue for the industrial dynamo and the intellectual vogue for thermodynamic and other energetic scientisms elicited from Adams the following lament:

> Nothing in the history of philosophy is more distinctly marked than the effort of physics and metaphysics, since 1890, to approach each other. . . . The naturalist now readily admits that plants have souls—or will-power,—but he appropriates the soul as an energy of thermodynamics. At first sight, the tendency seems towards metaphysics, but the true current is the reverse. The chaos is more chaotic than ever, but the effort to make the laws of *Energetik* cover all, is perhaps the only very vigorous intellectual activity now in evidence.[32]

Adams was alluding directly to Wilhelm Ostwald's theories of social energetics, in which the dictates of thermodynamics were applied to all other matters, and energy became the transcendental signified—the source and origin—of all natural and cultural dynamics.[33] As we have seen, the positivist extension of the second law of thermodynamics under the phrases "mechanical dissipation" and "increase of entropy" had also developed in visionary fashion from an observation about closed physical systems to a moral judgment on the material universe. Henry Adams's *Letter to American Teachers of History* documented the garish predictions of a number of venerable scientific pundits who extrapolated from the mechanical dissipation of energy to a near future in which all life becomes extinct in a final age of endless ice. The sun itself was seen to be dying away, already vastly contracted from its presumed former magnitude when the world was vigorous and carboniferous ferns grew on land now under ice caps. Adams cited this theory of "solar shrinkage" from "the standard authority in geology,"[34] Lapparent's three-volume *Treatise* of 1906, and then gave Lapparent's forecast for the cosmos. "Not yet satisfied," Adams remarked, "with admitting that organic geology, like inorganic, confirms the dissipation of energy down to the present day, M. Lapparent, abandoning all hope that the process can ever be reversed, concludes,"

> If any new term is still to be looked for, it seems as though none could be imagined other than an era where the Soul, freed from the bonds of matter, should dominate. Except for this hope there are none but sombre perspectives in sight for all that surrounds us. The progress of the emersion of boreal lands seems destined to extend from step to step the influence of the polar ice. The sun, whose condensation is already far advanced, will soon find in the narrowing of its diameter no sufficient source for maintaining its heat, and large spots will appear on its surface, destined to transform themselves into a dark shell. The day when the extinction of the central luminary shall be complete, no further physical or physiological reaction can take place on our globe, which will then be reduced to the temperature of space, and the sole light of the stars.[35]

In this instance, the extension of thermodynamic considerations from physical to geological discourse negated evolutionary appeals to the unforeseen origins and higher development of species in favor of an entropic vista of the sun's final extinction. Adams noted the irony in the grim sublimity of these scientifically authorized scientisms but apparently was unable to glean any solace from Lapparent's loophole, the possibility that the human soul might at some point in time rise beyond the contingencies of a dying body in an entropic world. With an irreversible temporal flow of entropy through a closed universe toward heat death, Victorian thermodynamic scientism gave a new urgency to the riddle of time and the desire for immortality, to the imaginative demand for some form of escape from time's material constraints.

One way to counter the dire visions of thermodynamic scientism was by recourse to another significant modern scientism, the doctrine of vitalism—the idea that life is fundamentally irreducible to material causes.[36] Operating in response to the dominant positivist forms of classical scientism, for several centuries vitalism developed as a broad romantic reaction against the mechanistic reductionism of the Newtonian paradigm. In medical, physiological, philosophical, poetic, and spiritualistic forms, vitalism and its counterscientisms flourished in the gaps left open by the program of classical physics. For most of its history vitalism operated within the pale of mainstream chemistry and biology as a valid countertheory to materialist mechanism. The cultural history of vitalistic ideas is a good example of the way that critical scientisms can be a form of ideological defense, a resistance to the violence of partisan scientific dogmatizing itself. In particular, vitalisms have been part of a complex response to the intellectual and affective threats posed by rigid Newtonian certainties and have expressed accurate intuitions of the very real limitations of classical scientific paradigms.

At the same time, vitalism's phantasmagoric remythologizing of nature has also impelled a long parade of cultural wish images. Conducted by its master trope—the *life force* that drives living things up the escalator of evolution just as the force of gravity pulls Newton's apple down—the energetic vitalisms of the era of classical thermodynamics conveyed scientific aura into a number of metaphysical and progressivist discourses. For instance, in 1889 British poet and social activist Edward Carpenter made a plea for the spiritual as well as medical significance of nudism, "sun-worship," a plea that coupled energetics and evolution to vitalistic notions of "health and radiance": "the process of evolution or exfoliation itself is nothing but a continual unclothing by Nature. . . . The consequent instreaming of energy into [Man's] system will carry him to perfections of health and radiance of being at present unsuspected."[37]

Adams's *Letter* invoked a more melancholy and defensive formation of vitalist scientism. Adams placed William Thomson's apocalyptic reading of the second law of thermodynamics against John Tyndall's *Heat as a Mode of Motion* of 1862 and remarked that Tyndall's grand concluding hymn to the first law's infinite conservation of energies buried in silence the dissipative outcomes of the second law. Caught between Tyndall's vision of the inexhaustibility of energy and Thomson's guarantee of its inexorable waste, Adams recalls, "the unscientific student could not interfere. Naturally, all his sympathies were with Tyndall."[38] The immediate continuation of Adams's commentary articulates the striking intellectual impasse of a modern humanist trying to negotiate the impossible positivist-scientistic imperative to place his own, discursive discipline—history—entirely under the conceptual rule of physics, in order to justify one's academic method as "scientific." Addressing himself in the third person, Adams

> was clear that the energy with which history had to deal could not be reduced directly to a mechanical or physico-chemical process. He was therefore obliged either to deny that social energy was an energy at all; or to assert that it was an energy independent of physical laws. Yet how could he deny that social energy was a true form of energy when he had no reason for existence, as professor, except to describe and discuss its acts? He could neither doubt nor dispute its existence without putting an end to his own; and therefore he was of necessity a Vitalist, or adherent of the doctrine that Vital Energy was independent of mechanical law.[39]

But this "necessary Vitalism" is clearly a defensive position within a cultural debate in which Adams has already virtually capitulated to the hegemony of positivist energetics. Adams articulated but could not resolve the double binds of an energic scientism that demanded the reconceptualization of all material and cultural substances as forms of energy. Thus for him the integral object of historical research had became a nebulous "social energy"; the phenomena of life were now expressions not of the quaint "life force" of Romantic physiology but of a "Vital Energy," which was still granted the paradoxical "meta-physical" character of a vitalistic agency. But if everything cultural is a form of energy, then all human relations are zero-sum conflicts leading to an inevitable inertia: a fight to the death for the possession of differentials, advantages, stockpiles, resources that hastens the ultimate waste land of total dissipation. With a century of hindsight we can now see that energetics and vitalism reached a mutual impasse in the era of classical thermodynamics because of the absence of an adequate discourse of information—that, as Gregory Bateson insisted, physical energy is a seductive but ultimately unworkable and often detrimental metaphor for biological life as well as for psychosocial processes.

Modernist Scientisms

In the era of classical thermodynamics, secular-positivist and theological-absolutist scientisms were jostled by further ranges of visionary, occult, or otherwise marginal or marginalized scientisms. For instance, the theory of the luminiferous and electromagnetic ether elaborated in Victorian physics impelled a wide array of visionary scientisms. Current notions of the cultural field as a charged matrix of discursive communications resurrect in an informatic idiom the earlier function of the ether as a cosmic energic medium, a subtle communicative fluid. This residue of the ether within the postmodern cultural field indicates the power of this past milieu of visionary scientism to mediate between matter and the immaterial, between wave energies and spiritual powers. The luminiferous ether was grasped in and out of science as a universal memory in which light waves not only transfer energies but also preserve signals. In this fashion Victorian physics set the stage for a general culture of media.

We will be especially interested in the modern cultural circuit between authorized and marginal technoscientisms. Visionary scientisms were common in modernist aesthetic discourse, often in the oppositional space of the little magazine and the journal of opinion. For instance, in her writings for the *Freewoman,* the *New Freewoman,* and the *Egoist,* British editor and essayist Dora Marsden articulated many of the ideological and formal premises of early modernism through a form of philosophical scientism derived from ether physics—the wave theory of radiant energy, and from neo-Lamarckian biology—a form of evolutionary vitalism. In the leader for the *Freewoman* of February 1, 1912, entitled "The Over-and-Above in Life," Marsden asserted that the "Over-and-Above,—Joy,—the Thrill,—this is the peculiar and single possession which makes life worth living."[40] She described the joy one feels in moments of supreme self-abandon: "in religious intoxication, in the passion for causes, in martyrdom, the joy of living is so intense and life so vivid that only the highest form of human love-passion can be compared with them." The desire for joy, Marsden claimed, is a spiritual motive that overrides all material impediments. But, she asks, "what is Joy? Who knows?" Her recourse was to define joy phenomenologically: "What each does know is that when he is joyous a pulse beats somewhere which creates and sets free an additional energy" (201).

In these phrases Marsden operated a metaphor between life and energy common in the modernist period, an electrovitalist conception of life as a polarized current adapted to the manifestations of physical energy in late-classical versions of thermodynamics and electromagnetism. She developed the sense of the "beating pulse" of ecstasy further as she shifted from the psychology to the *physics* of joy:

> Everywhere there appears to be this instinctive seeking towards vibration. So universal does it seem, and so essentially one in kind does its realisation appear to be, that the suggestion is forced on us that it is merely a specialised and self-conscious form of the rhythmic nature which is common throughout physical creation. What more do we know of light than that it is a form of vibration? Or of sound? Of ether, of electricity, or radium? May not life itself

be vibrant energy made self-conscious? And may not the instinctive desire for self-abandonment, the impulse to throw off restraint and surrender consciousness to the workings of an outside force, be a driving desire to implicate consciousness in the vibrations which we have become aware of in the physical world?[41]

This passage connects an existential quest for supraindividual being to Victorian physics and chemistry as reflected in theories of the luminiferous ether. The life force as an energy form often merged with the energy-conveying aura of the ether to produce the electrovitalism of Marsden as well as her contemporaries such as Ezra Pound and D. H. Lawrence. Marsden's identification of the primal pulse of creation with the "life force"— "Joy is the Life-Force, and the Life-Force is Joy"[42]—links the physics of wave theory to the progressive evolutionism of mystic vitalism. If those locked in radical struggle with the dead weight of oppressive social customs are energized by joy, and joy is the life force, then the life force itself must drive humanity into those political actions meant to effect the development of ever-higher living forms. As a result, says Marsden, "life learns at once its science and its religion" (202).

With regard to actual scientific developments in 1912—the identification of the primary atomic particles, the special theory of relativity—Marsden's vitalist rendition of wave theory is multiply poised on the edge of anachronism. But that is the common fate of the allegories constructed by cultural scientisms. Technoscientisms extend from the local sites of rapid scientific awareness to the larger and slower zones of broad cultural memory. What Marsden articulated so powerfully are the existential stakes of these particular scientific ideas in the early modernist milieu. Her discussion clarifies the vitalist essentialism in modernist evolutionism, by showing how strong the drive was to identify oneself with nature when nature itself could also be conceived not as a dying world but as a vibratory field in which life and mind are simply higher powers of the infinite energic undulations that underlie all creation. Marsden's radical social gospel is a profound example of the visionary scientism that animated much of the aesthetic militancy at the climax of the era of classical thermodynamics.

Ursula and the Physics Professor

D. H. Lawrence's writings epitomize the aesthetic militancy of early modernism. In addition, they dramatize the confluence and clash of positivist and visionary scientisms, in both energetic and vitalist modes, making mythopoetic narratives out of the tensions between science and religion. Toward the end of Lawrence's second major novel, *The Rainbow,* published in 1915, when the youthful heroine Ursula Brangwen returns to college to complete her university training, for the first year she dwells in the glamor of knowledge. She dreads her class in mathematics but enjoys those in botany. Then, in her second year, Ursula's attitude toward her professors shifts from reverential to disaffected, and like many idealistic college seniors torn apart by the possibilities of continued self-exploration on the one hand and the demands of making a living on the other, she yearns for some form of ecstatic resolution. So, she switches majors: "She would take honors in botany. . . . She had entered into the lives of the plants. She was fascinated by the strange laws of the vegetable world. She had here a glimpse of something working entirely apart from the purpose of the human world."[43] Botany here is a special field of scientific investigation with "strange laws" that seem to intimate something beyond *human* purposes, that seem to defy the bourgeois commodification of the cosmos purveyed, as we saw in the last chapter, in the discourse of Victorian thermodynamics. When Ursula merges with the plants' life, she also enters into one of *The Rainbow*'s central metaphors for spiritual growth: the inner self or soul is a germinating seed or an unfolding bud, and spiritual progress is the development or decay of that seed, the blossoming or blighting of that bud. Such spiritual botany captures Lawrence's wider conceptual investment in Life as a source of values and, as it is for Ursula, a "source of mystery" (436).

One day her thoughts wander to a recent conversation with her physics professor, one not previously depicted, and apart from Ursula's recollection we never actually meet this female professor with the ominous name of Dr. Frankstone. It is implied that Ursula had approached Dr. Frankstone to venture a hypothesis about the "strange laws" of life, for she is now "fretting" over the scientist's materialist rebuff of her vitalistic desire for some form of "special mystery": "'No, really,' Dr. Frankstone had said, 'I don't see why we should attribute some special mystery to life—do you? . . . May it not be that life consists in a complexity of physical and chemical activities, of the same order as the activities we already know in science? . . .' The conversation had ended on a note of uncertainty, indefinite, wistful."[44] This scene stages a stark confrontation between classical determinism and romantic vitalism.

In search of purpose and direction, Ursula feels multiply threatened. The everyday world strikes her as a nullity, and science, too, seems to shut the door on her quest for special exemption from the ordinary, seemingly lifeless and pointless properties of matter and energy. Ostensibly we have a scientific debate between Ursula's vitalism and Dr. Frankstone's materialistic physics, or in the parlance of late Victorian science, the "atomo-mechanical theory."[45] Although Lawrence means our sympathies to veer away from the dogmatic monster of dead matter Frankstone and fix on Ursula's spirit of living mystery, contemporary readers may reflect that molecular biology has borne out Dr. Frankstone's position with regard to the mechanisms, if not the wonder, of life and eliminated the last vestiges of scientific plausibility that Ursula's implied claims still possessed in Lawrence's day.

But what really bears attention in this passage is the theological tenor of Ursula's response to the scientist's doctrines. Ursula's vitalism does not really rest on any specifically botanical or biological grounds: it is not really an alternative science but rather a visionary scientism. Just as science had presumed to dictate the terms of reality to her, she will presume to defy them and invent her own. Like Lawrence himself, certain of his characters "lose their science," just as one might otherwise lose one's religion. This reflects the way in which an important portion of modernist literature defined itself by breaking away from Victorian positivism and its "religion of science." In Lawrence's case, the positivist scientisms of the educational establishment are countered not just by primitivistic doctrines or primal emotions, but also by visionary scientisms in which the premises of authorized science are imaginatively extended beyond the limits of empirical validation.

As she continues to brood over the physicist's remarks, then, Ursula redirects the argument to a position beyond the standard limits of science by introducing philosophical and religious issues of purpose, will, and soul. For in her afterthoughts, she now replies to Dr. Frankstone: "the purpose, what was the purpose?"

> Electricity had no soul, light and heat had no soul. Was she herself an impersonal force, or conjunction of forces, like one of these? She looked still at the unicellular shadow that lay within the field of light, under her microscope. It was alive. . . . What then was its will? If it was a conjunction of forces, physical and chemical, what held these forces unified, and for what purpose were they unified?[46]

Ursula resists Dr. Frankstone's desacralized world by demanding ultimate answers to the sort of teleological questions that modern science disavows. Physicists are simply not accustomed to conceiving of particles or waves in terms of their spiritual goals. And yet because modern science by Ursula's day had already achieved such comprehensive social authority, it inevitably becomes culturally implicated in her processes of existential orientation. Ursula's search for her own individual purpose, then, is projected through the apparatus of science—the microscope—onto the living macrocosm, as epitomized in the minute microcosm of her illuminated slide. In Ursula's militant biological allegory, Life wills not just to exist, but also to fulfill itself, to be all that it can be. Still peering at the microbe, her ruminations continue:

> It intended to be itself. But what self? Suddenly in her mind the world gleamed strangely, with an intense light, like the nucleus of the creature under the microscope. Suddenly she had passed away into an intensely-gleaming light of knowledge. She could not understand what it all was. She only knew that it was not limited mechanical energy, nor mere purpose of self-preservation and self-assertion. It was a consummation, a being infinite. Self was a oneness with the infinite. To be oneself was a supreme, gleaming triumph of infinity.[47]

Ursula's unanswerable retort to Dr. Frankstone, then, takes shape in this illuminated moment of sublime self-reflection. Lawrence's larger retort to the stereotypical secular science parodied in the female physicist is to celebrate the supremacy of the self's potential infinitude—in short, the immortality of the soul—over the uncertainty and self-effacement demanded by science's objectivity and impersonality. As we will see in chapter 8, however, a few years later Lawrence replayed the confrontation between science and philosophy staged by this scene from *The Rainbow,* with some interesting variations, when in his own discursive person he encountered another, rather different, real-life physicist, Albert Einstein.

Revolution versus Entropy

> Surely you see that only differences, differences of temperature, only contrasts in degree of heat, only that makes for life? And if throughout the universe all bodies are equally warm, or equally cool . . . You've got to smash them into each other—so there'll be fire, explosion, inferno.
>
> —I-330 to D-503 in Zamyatin's *We*

The violent social impasse between entropic mechanism and evolutionary vitalism takes high-modernist form in Russian author Yevgeny Zamyatin. Along with sustained satires on energetics and eugenics, in *We* Zamyatin also spoofs the virtually Pythagorean mathematicism that operates in the dystopian OneState's milieu of deterministic positivity. *We*'s allegory of mathematical scientism mixes classical thermodynamics with the leading edge of Einsteinian physics. The novel is especially remarkable for its adaptation of mathematical concepts as metaphors for aesthetic, social, and political formations. As we will observe in chapter 8, it also culls many of its mathematical allegories from the main line of non-Euclidean geometries and hyperspace theories that formed the historical context for the reception of Einstein's relativistic cosmos.

But the base on which Zamyatin constructed his critiques as well as his own deployments of scientistic discourse was classical thermodynamics. A naval engineer by training, Zamyatin's take on thermodynamics was forged in part by the development of steam technology applied to marine vessels.[48] In any event, both the first and second laws of thermodynamics animate Zamyatin's vigorous 1923 essay, "On Literature, Revolution, Entropy, and Other Matters." This polemical tract elucidates the thematic construction of *We* while providing a synopsis of critical scientisms that circulated throughout the Victorian and modernist periods, transferring the aura of physical energies to a mixture of aesthetic and social doctrines. Here Zamyatin factored the revolutionary transformation of Russia into the Soviet Union directly into thermodynamic concepts: "the law of revolution is not a social law, but an immeasurably greater one. It is a cosmic, universal law—like the laws of the conservation of energy and of the dissipation of energy."[49] Thus the onset of the Soviet regime is only a catalyst for, not a cessation of, further upheavals. Zamyatin also granted revolutionary status to the Russian mathematician Lobachevsky's development of hyperbolic geometry in the preceding century, and hailed Einstein's adoption of non-Euclidean geometry to field physics as the latest phase of this larger transformation: "Euclid's world is very simple, and Einstein's world is very difficult—but it is no longer possible to return to Euclid." Zamyatin recuperated the cultural agonies of his time by reading revolutionary intellectual and political wounds as medical interventions for a dissipated or "fatigued" society: "most of mankind suffers from hereditary sleeping sickness, and victims of this sickness (entropy) must not be allowed to sleep, or it will be their final sleep, death."[50]

So Zamyatin contrasted the "law" of revolution with the social trope of entropy as the somnolent inertia of an exhausted culture: "The law of revolution is red, fiery, deadly; but this death means the birth of new life, a new star. And the law of entropy is cold, ice blue, like the icy interplanetary infinities."[51] Under the regime of these critical scientisms, the gradual degradation of cultural energies seems as starkly certain as the drift of thermal energies toward exhausted equilibrium: "When the flaming, seething sphere (in science, religion, social life, art) cools, the fiery magma becomes coated with dogma—a rigid, ossified, motionless crust. Dogmatization in science, religion, social life, or art is the entropy of thought" (108). As developed at length in *We,* Zamyatin proposed to arrest and reverse the devolutionary cultural slackening toward maximal conformist entropy through sudden blasts of liberatory violence.

In these energized metaphors, "revolution" comes to stand for any physical or cultural phenomenon that can be credited as both "energetic" and "counterentropic." Zamyatin's thermodynamic scientism awarded doctrinal preference specifically according to the perception of counterentropic forces—the sorts of cultural vitality that reverse the entropic fatigue of social conformity. He located these forces at the discursive level in ever-renewed "heresies": briskly restating early-modernist artistic doctrines, Zamyatin advocated a perpetual creative defiance of established cultural dogmas. He figured this combination of aesthetic and social agendas at the cosmic level through the image of a violent transformation of burned-out worlds, creative crashes into fiery life. Zamyatin's thermodynamic allegory of social renovation through destruction is epitomized in the following key formulation: "Two dead, dark stars collide with an inaudible, deafening crash and light a new star: this is revolution."[52]

Zamyatin's scientific emblem for creative modernist catastrophe is also an apt image of the destruction piled up before the angel of history's startled gaze by the global debacles of the first half of the twentieth century. The collision of thermodynamic and vitalist scientisms led to a peculiar impasse that significantly aggravated that intellectual milieu. The closed-systematics of classical thermodynamics created a scientistic feedback loop, a Red Spot in the turbulence of a newly planetary culture, a rhetorical bias in the social operation of the cultural field that produced an impressive amount of creative response and critical resistance among writers and artists, but that also favored the political organization of OneStates, closed totalitarian societies, as a compelling but clearly futile hedge against "entropy." In part 2, moreover, we will discover that the collision of dark stars is an idea deeply embedded in the earliest phases of the era of classical thermodynamics.

In sum, we have asserted that technoscience has long been complicit with the cultural politics of allegorical coding. The sciences' own aspirations to cultural authority have driven the modern phantasmagorias of technoscientism. Technoscientism comes into conceptual relief as science arrives at its institutional majority. The ascendancy of technoscience as a dominant cultural discourse, its accumulation of social capital within the field of cultural power, has generated its social commodification in the rhetorical forms of scientism. Allegorical codes have been increasingly displaced from religious duties and redeployed in the service of the new, scientific forms of cultural authority. Just as the concept of technoscientism provides a social and historiographical model for many literary and artistic productions, the discourse of allegory provides formal and thematic models for scientistic productions. Allegorical codings are implicit in the sociopolitical formations of scientism and become increasingly explicit in its creative formations. The allegorical functions of scientism establish a cultural mediation between skeptical positivism and the visionary extension of the sciences beyond the range of the empirical. The allegorical function of scientism also allows belief to be carried over to the undemonstrable on faith in scientific, as opposed to religious or customary, authority.[53] But the allegorical forms and functions of scientism also operate within the fissures of cultural belief-structures. The heuristic invention of imaginable yet unconfirmed objects in scientific and artistic productions may undermine as well as reconfirm the stability of prior scientific paradigms and their cultural authority.

PART 2

ALLEGORIES OF THERMODYNAMICS

Chapter 4
Maxwell's Daemonic Science

> The sciences themselves know of those angels whom you yourself have elsewhere called "delegates," or envoys, sent to observe more subtly than we can.
>
> —Michel Serres to Bruno Latour

Maxwell in *The Crying of Lot 49*

As an intermediary discourse, allegory has always been a shifty enterprise, vacillating between desire and constraint, earnest dogmatism and satirical revisionism. This oscillation is moral as well as modal. In allegory the human and the divine regularly collapse into their ambivalent middle term—the daemonic, the indeterminate and unpredictable realm of cosmic mediation. In other words, the messages that circulate within allegories are interrupted by their own media. In these narratives, social effects emerge from semiotic deviations—uncanny or disruptive supplements, the noise or excess infecting or reorganizing transmissions.[1]

The daemonic figures native to mythic landscapes personify allegory's historical functions of cultural mediation. Daemons are agents of communication, typically taking the shape of messengers, or guardians, or other figures of admonition. The crucial thing is that they can take whatever shape the larger conceptual scheme demands: they are inherently metamorphic. From the metamorphs of the classical tradition and the angels and devils of Judeo-Christianity to the robots, aliens, and (d)evolved beings of science fiction and the time-traveling cyborgs of contemporary action cinema, fictive agents perennially emerge at allegorical thresholds, bridging conceptual gaps and bearing important cultural information. The scientific daemonic in particular is an important mythopoetic register of technoscientism. The angelic and daemonic agents in allegories of energy are winged runners in a cosmos technoscience has redescribed as a global communication matrix.[2]

Modern science has produced its share of daemonic figures.[3] As we have previously remarked, in the period preceding the emergence of thermodynamics, the French mathematician and astronomer Laplace (1749–1827) invented a Demon to characterize the ultimate intellectual power conceivable through the determinism of Newtonian dynamics. Aware at any given instant of the position and velocity of every material particle in the universe, Laplace's Demon possessed a perfect calculus of the past and the future.[4] Similarly, following the formulations of Carnot, Helmholtz, Thomson, Clausius, and Maxwell, the laws of thermodynamics gave rise to a complex, often contradictory range of social ideologies and creative responses personified by daemonic figures.[5] At times the same scientists developing the physical theories also initiated the allegorization of thermodynamics by packaging the physical implications of energy concepts in strong rhetorical creations. Perhaps the most famous of all the rhetorical daemons in modern science is the "neat-fingered being" invented by British physicist James Clerk Maxwell.

We will look closely at the specific origin and function of Maxwell's Demon, but at the moment I want to sketch some of its significations within the larger allegory of classical thermodynamics in Victorian culture. As a messenger from science, this theoretical daemon arrived to speak to William Thomson's deterministic reading of the second law of thermodynamics, in which the mechanical energy of a closed system must invariably become perfectly dissipated and hence useless for human purposes. Our universe, too, it was thought, must ultimately be such a closed system, so it must also be approaching an inevitable heat death. To escape the fate of such mechanical dissipation, Maxwell granted his Demon the microscopic perceptual and manipulative faculties to sort slower and quicker molecules into separate vessels, "without expenditure of work . . . in contradiction to the second law of thermodynamics."[6] This sorting operation virtually reversed time, allowing dissipated heat to fall once more and thus, if coupled to the proper engine, to produce further mechanical work. The Demon's intelligent touch thus counteracted the mindless cosmic drift toward thermal equilibrium.

Maxwell's Demon is a quite specific structural instance of a daemonic agent, what Angus Fletcher has shown to be the generic form of allegorical agency: "The allegorical hero . . . conforms to the type of behavior manifested by people who are thought (however unscientifically) to be possessed by a daemon."[7] And the daemon of diffuse anxiety that possessed Maxwell was entropy, the vision of evil encapsulated in the Victorian thermodynamic prophecy of imminent heat death. The genius of Maxwell's Demon was the desire to secure conceptual salvation from the finality of that last judgment. Like any agent of spiritual or philosophical allegory, Maxwell's Demon was conceived in practical response to a moral apprehension of cosmic totality.[8]

However, the demon quickly exceeded its original occasion, going on as a potent conceptual operator in physics and computer science to promote advances in statistical mechanics, the analysis of probabilities, and the ongoing redescription of physics and biology in cybernetics and information theory. In their summary of the history of Maxwell's Demon as a conceptual operator in the hard sciences, Harvey S. Leff and Andrew F. Rex underscore its longevity despite repeated efforts to evacuate its claims to model plausible physical behaviors.[9] From our perspective, the Demon's hardiness is due exactly to its allegorical adequacy as an adaptable model of physical and informatic systems. Despite their daemonic status, "allegorical agents are *real enough,* however ideal their referents may be, however 'unlike ourselves' they may appear. They have what may be called an 'adequate representational power.' "[10]

Thomas Pynchon's *The Crying of Lot 49* commemorates and updates the combination of the daemonic and the thermodynamic in the creation of Maxwell's Demon. This postmodern quest-allegory presents daemonic phenomena in several guises—as material communications, in the Thurn and Taxis monopoly and W.A.S.T.E. system of outlaw mail delivery; and as psychic possession, in Oedipa Maas's free-floating paranoia concerning the reality of the Trystero.[11] After she sees the threshold vista of San Narciso as a hieroglyph for the printed circuits of communications devices,[12] Oedipa's world remains enigmatic, emblematical. Messages from unknown, perhaps hallucinated, correspondents force her to acknowledge the possibility of entirely different levels of psychical or cosmic determination. She lives inside a web of dark conceits beyond her grasp. *The Crying of Lot 49* places the sign of Maxwell's Demon over a novel that extends the formulary analogy between thermodynamic and information entropy into a mock-apocalyptic social allegory. Mathematical analogy becomes scientific ideology.[13]

Midway through the text, Yoyodyne engineer Stanley Koteks turns Oedipa Maas on to the existence of the Nefastis Machine—Pynchon's postmodern spoof of nineteenth-century perpetual-motion engines and of their theological analogues, the ghostly claims of various Victorian spiritualists.[14] In a sort of séance with the device, the operator of the Nefastis Machine uses telepathy with a Maxwell's Demon to finesse the mechanical dissipation mandated by the second law of thermodynamics:

> From a drawer he produced a Xeroxed wad of papers, showing a box with a sketch of a bearded Victorian on its outside, and coming out of the top two pistons attached to a crankshaft and flywheel.
>
> "Who's that with the beard?" asked Oedipa. James Clerk Maxwell, explained Koteks, a famous Scotch scientist who had once postulated a tiny intelligence, known as Maxwell's Demon . . . . He went on to tell how the

> Nefastis Machine contained an honest-to-God Maxwell's Demon. All you had to do was stare at the photo of Clerk Maxwell, and concentrate on which cylinder, right or left, you wanted the Demon to raise the temperature in. . . . The familiar Society for the Propagation of Christian Knowledge photo, showing Maxwell in right profile [fig. 11], seemed to work best.[15]

The demon's position at a mediating threshold between distinct spaces conforms to legendary traditions. Transporting Maxwell's Demon from its Victorian origin to the modern conceptual threshold between the sciences of energy and information by invoking the analogy between Boltzmann's statistical equation for thermodynamic entropy and Shannon's probabilistic equation for information-entropy[16]—Nefastis tries to explain to Oedipa how the dual dynamisms of entropy drive his machine. It operates, says Nefastis, by the literalization of a metaphor:

> "Help," said Oedipa, "you're not reaching me."
>
> "Entropy is a figure of speech, then," sighed Nefastis, "a metaphor. It connects the world of thermodynamics to the world of information flow. The Machine uses both. The Demon makes the metaphor not only verbally graceful, but also objectively true."
>
> "But what," she felt like some kind of heretic, "if the Demon exists only because the two equations look alike? Because of the metaphor?"
>
> Nefastis smiled; imperturbable, calm, a believer. "He existed for Clerk Maxwell long before the days of the metaphor."
>
> But had Clerk Maxwell been such a fanatic about his Demon's reality?[17]

The undetermined reality of the demon is precisely to the point. Nefastis is parodically un-Maxwellian, if not un-Victorian, in giving the demon a literal existence. Maxwell's own attitude toward his physical models was more skeptical. The success of his science depended in part on his rhetorical sophistication in allowing hypothetical models to remain metaphors.[18] At the same time, Nefastis is right: despite efforts to exorcise it, the demon still possesses virtual existence as an active scientific allegory. In fact, it was William Thomson, later Lord Kelvin, who began this tradition of scientific reification by naming as a "demon" the fictive intelligence that Maxwell had conceived in 1867 simply as a "neat-fingered being." Maxwell was not entirely happy with this stark anthropomorphism, and in subsequent correspondence with others he attempted to demote the creature from intelligent to mechanical action. The simplicity of its function "reduces the demon to a valve," he explained to Tait in an undated note; in an 1870 letter to John William Strutt the demon is "a mere guiding agent (like a pointsman on a railway with perfectly acting switches . . .). I do not see why even intelligence might not be dispensed with and the thing made self-acting."[19]

Fig. 11. Maxwell in right profile. Frontispiece of *Matter and Motion* (1876).

Despite Maxwell's skepticism, Thomson's name "demon" was an inspired rhetorical choice, one that has possessed poor Maxwell ever since.

Maxwell's Poetic Works

As Maxwell was instrumental in coding allegory into the discourse of Victorian thermodynamics, we will dwell for a while on his methodological development. Much has been made of Maxwell's exposure while studying in Edinburgh to the Common Sense school of Scottish philosophy, a doctrine that emphasized concreteness, economy, and caution.[20] I want to suggest that these components of Maxwell's philosophical culture were both enhanced and counterbalanced by his simultaneous immersion in a particular literary culture—a Anglo-Scottish hermeneutic tradition compounded of Protestant moral allegory and Romantic aesthetics—which also had a significant effect on his scientific style and considerable achievements. C. W. F. Everitt makes the point that "many of the early signs point to a flowering in literature rather than science"; Maxwell received a strong classical education at Edinburgh Academy and was at the top of his literature classes.[21] Important aspects of Maxwell's scientific methodology may be seen as a form of *allegoricism* extending to scientific practice the techniques of allegorical reading and writing.

Maxwell left a modest body of light poetic work. His lesser poems are mostly school exercises, but some of his best are satiric verse commentaries on the scientific conference scene, in particular, epitomes or send-ups of papers delivered at the annual meetings of the British Association for the Advancement of Science. In chapter 1 we noted excerpts from Maxwell's verse report on Tait's lecture on force. Similarly, in a letter to Tait after the 1874 meeting of the British Society, Maxwell penned an epigram on "domestic evolution" skewering Herbert Spencer's evolutionistic address to the convention:

> The ancients made enemies saved from the slaughter
> Into hewers of wood and drawers of water;
> We moderns, reversing arrangements so rude,
> Prefer ewers of water and drawers of wood.[22]

Maxwell's school poems lack the terse pungency of his later parodies, but some of them bear up under examination. One is "A Vision," written in his third year at Cambridge in chiming trochaic stanzas reminiscent of Poe's "The Raven":

> Deep St. Mary's bell had sounded,
> And the twelve notes gently rounded
> Endless chimneys that surrounded
> My abode in Trinity.
> (Letter G, Old Court, South Attics),
> I shut up my mathematics,
> That confounded hydrostatics—
> Sink it in the deepest sea![23]

The student-poet reflected on the seeming futility of his curriculum:

> In the grate the flickering embers
> Served to show how dull November's
> Fogs had stamped my torpid members,
> Like a plucked and skinny goose.
> And as I prepared for bed, I
> Asked myself with voice unsteady,

If of all the stuff I read, I
Ever made the slightest use.

To answer his dejection over grinding away at obscure mathematical topics in preparation for the tripos, a weeklong exam administered in the winter of a Cambridge student's fourth year, a first vision arises: "Look beyond, and see the wrangler, / Now become a College dangler." As Martin Goldman reports, "to get a first in the tripos, to become a wrangler, was to get a meal ticket for life."[24] But this vision of future ease does not console the poet, for the imaginary wrangler, smitten with "worldly Pride," foregoes "restive Reason" and turns into a inert conservative: "See him next with place and pension, / And the very best intention / Of upholding that Convention / Under which his fortunes rose."

So the poet finds no immediate solace in his vision of the worldly wrangler for whom academic success has become an opportunity for intellectual laziness and institutional conformity. His reverie continues:

Still those papers lay before me—
Problems made express to bore me,
When a silent change came o'er me,
In my hard uneasy chair.
Fire and fog, and candle faded,
Spectral forms the room invaded,
Little creatures, that paraded
On the problems lying there.

With the benefit of hindsight we might consider these "little creatures" to be an early influx of "Maxwell's demons," fictive beings that emerge from the imaginative consideration of knotty intellectual problems. However, at the moment these particular parading "creatures" form a panorama of the university in its perpetual cycle of drill and examination, its pomp of eventual graduation, and "All the costly apparatus, / That is meant to elevate us / To the intellectual status / Necessary for degrees."

The poet's confession of disdain for all these sanctified characters of entrenched Cambridge tradition now provokes an ironic figure of righteous admonition to descend upon him.

Angular of form and feature,
Unlike any earthly creature,
She had properties to meet your
Eye whatever you might view.

Hair of pens and skin of paper;
Breath, not breath but chemic vapour;
Dress,—such dress as College Draper
Fashions with precision due.

Not as the Lady Philosophy had come down to soothe the torment of Boethius, here the "awful form" of Pedantry materializes to upbraid the poet for his "unsanctioned" displeasures. But also, as an alma mater figure, Pedantry attempts to lead her prodigal charge back into conformity with proper Cambridge ambitions.

"Powers!" she cried, with hoarse devotion,
"Give my son the clearest notion
How to compass sure promotion,
    And take care of Number One.
Let his college course be pleasant,
Let him ever, as at present,
Seem to have read what he hasn't,
    And to do what can't be done."

The seriocomic climax of this visionary admonition to persevere toward the tripos concerns the practical choice between poetic and scientific vocations. Pedantry's prayer concludes:

Of the Philosophic Spirit,
Richly may my son inherit;
As for Poetry, inter it
    With the myths of other days.
Cut the thing entirely, lest yon
College Don should put the question,
Why not stick with what you're best on?
    Mathematics always pays.

In fact, Maxwell eventually became second wrangler and assumed a fellowship at Trinity College. But the confrontation he staged between Pedantry and Poetry—"calculation" and "worship"—expressed an important distinction between secular erudition and religious awe, to which the poet turns in the last movement of the poem. Now a final vision emerges of a veiled female figure whose features are left in sublime obscurity, but whose name is finally uttered—Nature, the true object of "natural philosophy":

I could never finish telling
You of her that has her dwelling
Where those springs of truth are welling,
  Whence all streams of beauty run.
She has taught me that creation
Bears the test of calculation,
But that Man forgets his station
  If he stops when that is done.

Is our algebra the measure
Of that unexhausted treasure
That affords the purest pleasure,
  Ever found when it is sought?
Let us rather, realising
The conclusions thence arising
Nature more than symbols prizing,
  Learn to worship as we ought.

In this Wordsworthian ode, mere Pedantry is allied to a nominalistic or self-centered concentration on mathematical "symbols." In contrast, Poetry bespeaks the glory of a philosophical Nature that calmly yields herself to the calculative measurements of "restive Reason," but whose sources remain ultimately metaphysical, beyond calculation's reach. Maxwell is culturally located within the tradition of natural theology, an intellectual current that ran powerfully through British physics of the nineteenth century, and that also found avatars in Maxwell's scientific colleagues William Thomson, Balfour Stewart, and Peter Guthrie Tait. I have drawn attention to Maxwell's poetry, however, not to underscore its moral admirability or conventional piety, but to reflect on its easy and continued facility with a particular poetic device, the trope of personification. Maxwell's collegiate vision bodies forth his moral self-inventory with "little creatures" that coalesce on his desktop, with Pride and Pedantry wrestling Reason and Natural Philosophy for the poet's allegiance. A procession or concatenation of reified abstractions is a standard marker of doctrinal argument in literary and visual allegory.

Allegory and Analogy

Well prior to his encounter with the concept of entropy and invention of the Demon, Maxwell invoked the allegorical mode in some of his earliest scientific papers. Directly out of college, as a fellow at Trinity Maxwell began an intensive review of theories of electricity and magnetism. He wrote to William Thomson in February 1854, requesting advice on a course of study. On September 13, 1855, Maxwell wrote to Thomson again, to discuss the construction of his own theories:

> My object . . . was . . . to learn what had been done in electrical science, mathematical & experimental, and to try to comprehend the same in a rational manner by the aid of any notions I could screw into my head. In searching for these notions I have come upon some ready made, which I have appropriated. Of these are Faraday's theory of polarity . . . , also his general notions about "lines of force" with the "conducting power" of different media for them.
>
> Then comes your allegorical representation of the case of electrified bodies by means of conductors of heat. . . . Then Ampère's theory of *closed* galvanic circuits, then part of your allegory about incompressible elastic solids.[25]

The same letter goes to pronounce for a third time the term in question: "you published a fragment of your speculations in the form of an allegory about incompressible elastic solids" (1:322).

Maxwell's use of the term *allegory* here is unusual: the more common term is *analogy*. Maxwell's career would climax in his seminal theory of the electromagnetic field.[26] At the outset of that career he used *allegory* to think about the construction and unification of scientific theories, about the heuristic shifting of explanatory models from one order of phenomena to another. Among the things he was discussing with Thomson in 1855 was the latter's 1847 paper, "On a Mechanical Representation of Electric, Magnetic and Galvanic Forces," proposing "an analogy for the electric and magnetic forces by means of the *strain,* propagated through an elastic solid."[27] In later years Maxwell authored several papers on the question of scientific analogies. His analogical methods called for the mechanical modeling or concrete visualization of imponderable physical processes, and also for the investigation of the mathematical analogies that emerged among different physical systems:

> When the student has become acquainted with several different sciences, he finds that the mathematical processes and trains of reasoning in one science resemble those in another so much that his knowledge of the one science may be made a most useful help in the study of the other. When he examines into the reason of this, he finds that in the two sciences he has been dealing with systems of quantities, in which the mathematical forms of the relations of the quantities are the same in both systems, though the physical nature of the

quantities may be utterly different. He is thus led to recognize a classification of quantities on a new principle, according to which the physical nature of the quantity is subordinated to its mathematical form.[28]

Scientific discourse typically concerns conceptual abstractions stated in the language of mathematics. The allegoricism of Maxwell's physics, the total cultural circuit by which he transposed traditional verbal structures and rhetorical concepts from hermeneutic to experimental application, results in a sort of feedback loop from literary and religious training to scientific method. In addition, analogy and allegory share an affinity for algebra and geometry, for instance, the doctrinal applications of continued geometrical proportion we noted in Platonic cosmology. But relative to scientific analogy, the notion of allegory has even greater conceptual resonance, and it is reasonable to think that Maxwell's characterization of Thomson's scientific papers as "allegorical" was a moment of significant wordplay. Unlike analogy, which often denotes a fairly unambiguous and deliberate act of numerical or empirical comparison of formal, relational, or substantial resemblances, allegories implicitly invoke a series of Western discourses connecting literary and religious writings to cosmological and mechanistic schemes. Allegory is the rhetorical form most *like* mathematics in the symbolic equations that build conceptual models. Allegorical texts present a discursive algebra of reified images and emblems, a structural geometry of networked spaces.[29]

Set against his remarks on analogy and beside William Thomson's later naming of the demon, the odd usage of "allegory" in his early letter to Thomson shows Maxwell's own literary flair, cultured in a mid-Victorian Anglo-Scottish milieu that connected literature and philosophy to mediate between Presbyterian and Anglican persuasions. For Maxwell and his circle, physics gives mathematics a body of material constraints from which to transcribe abstract relationships and formulate physical laws, just as it gives to theology an object in which the ultimate nature of its creator can be cognized.[30] Smith and Wise relate that Thomson and Maxwell "shared a Scottish tradition in natural philosophy which stressed both the separation of mind from matter and the directing power of mind over matter."[31] Maxwell and many in his circle were Christian Platonists for whom Nature is an allegorical text reifying conceptual and spiritual truths, making them concrete and so accessible to analytical comprehension and rational control. Maxwell's Victorian physics came by allegory "naturally" in its aspirations simultaneously to mathematize and theologize the totality of world-systems.

Maxwell's Electromagnetic Ether

> Before Maxwell people conceived of physical reality—in so far as it is supposed to represent events in nature—as material points, whose changes exist exclusively of motions. . . . After Maxwell they conceived physical reality as represented by continuous fields, not mechanically explicable. . . . This change in the conception of reality is the most profound and fruitful one that has come to physics since Newton.
>
> —Albert Einstein, "Maxwell's Influence on the Evolution of the Idea of Physical Reality"

In the history of energy physics, Maxwell's contributions to thermodynamics are surpassed in importance by his theories of the electromagnetic field. At the outset of his work on electromagnetism, Maxwell investigated Faraday's field theories by extending the forms of mathematical analogies into heuristic models of constructed, reified effects. His first important paper on electromagnetic theory is "On Faraday's Lines of Force," delivered to the Cambridge Philosophical Society in 1855–56. "In order to obtain physical ideas without adopting a physical theory we must make ourselves familiar with the existence of physical analogies," Maxwell commented in the discursive introduction to the paper: "By a physical analogy I mean that partial similarity between the laws of one science and those of another which makes each of them illustrate the other."[32] In Maxwell's scientific method, mathematics provided the language for the recognition and statement of physical analogies. These analogies were suggested by similarities not in the functioning but in the *equations* of differing physical systems:

> There is no formula in applied mathematics more consistent with nature than the formula of attractions, and no theory better established in the minds of men than that of the action of bodies on one another at a distance. The laws of the conduction of heat in uniform media appear at first sight among the most different in their physical relations from those relating to attractions. The quantities which enter into them are *temperature, flow of heat, conductivity*. The word *force* is foreign to the subject. Yet we find that the mathematical laws of the uniform motion of heat in homogenous media are identical in form with those of attractions varying inversely as the square of the distance. We have only to substitute *source of heat* for *center of attraction, flow of heat* for *accelerating effect of attraction* at any point, and *temperature* for *potential,* and the solution of a problem in attractions is transformed into that of a problem in heat.[33]

For this method of physical analogies Maxwell credited the papers of William Thomson, in which heat flow and electromagnetism had been related by means of an analogical transfer in their mathematics, showing that "the equations of heat flow in an infinite solid apply equally well to static electricity, when the symbols are reinterpreted. The symbol for 'heat source' should be interpreted as 'charge' and that for 'temperature' as 'potential.' "[34] Here is the one-to-one transcoding of different physical "levels" with mathematical symbols as a bridge that Maxwell recognized as a form of figurative gambit transferred from hermeneutic to scientific investigation. Thus his references to Thomson's "allegories."

Just as the literary use of figurative expressions gives rise to interpretive questions about appropriate levels of literal conviction, the scientific method of physical analogies raises questions about the realism to be invested in heuristic models. Maxwell repeatedly emphasized that his physical analogies were not real descriptions but figurative aids to mechanical visualization. Like any literary trope, they were to be grasped as poetic vehicles that gestured toward an as-yet unascertained reality.[35] In "On Faraday's Lines of Force," Maxwell operated on the mathematical analogies in the equations of fluid flow and electricity by envisioning the lines of electromagnetic force as streamlines in layered tubes of an incompressible fluid. As with the paper by Thomson just discussed, "On Faraday's Lines of Force" performs a transcoding or heuristic substitution of physical components:

> The mechanical "analogy" for electricity looks like this. Positive charge is considered as a "source" of the fluid. The charge continually sends the fluid out at a rate depending on the intensity of the charge. Negative charge is considered as a "sink" which absorbs the fluid from all around at a rate depending upon its intensity. The amount of negative and positive charge is equal, according to Faraday, so the sinks never lack fluid to absorb. This is, indeed, a strange fluid which can be continuously created in one place and continuously destroyed in another. Maxwell emphasized that, since it is only a mathematical analogy to the phenomena, we can assign it whatever properties we wish.[36]

In our terms, the poetic component of Maxwell's analogical method here is the instigation of an allegorical interplay between the abstract and the concrete, the literal invisibility of the electrical phenomena under investigation and the figurative ponderability of the hydrodynamic analogue devised to model their workings. Maxwell noted in his summary of the first part of "On Faraday's Lines of Force" that "my aim has been to present the mathematical ideas to the mind in an embodied form."[37] " 'Embodied mathematics'—mathematics represented in mechanical examples" was "Maxwell's enduring commitment":[38] this locution brings us powerfully into the domain of allegorical reification. Yet Maxwell pursued his experimental allegories nondogmatically, not to inculcate a doctrine but to construct a testable hypothesis.

In the same month that Maxwell completed "On Faraday's Lines of Force"—February 1856—he delivered another paper at Cambridge, this time to the elite debating society known as the Apostles.[39] In "Are There Real Analogies in Nature?" Maxwell is in a relatively light discursive vein. His remarks circle around some philosophical questions that issue from the confrontation between the physical world and the intellect in their Kantian descriptions. The overall tone of the piece is suggestive rather than conclusive. The question about "real analogies" pertains to the possibility of actual—as opposed to mathematical or symbolic—congruence between the phenomenal operations of the human mind and the substantial world it takes as its object. Maxwell's locution of "real analogy" as a bridging term represents a philosophical stance of greater epistemological caution than, say, the claim of complete fusion between the world and our mental representations of it, but also of less epistemological skepticism than the claim of complete disjunction.

For instance, Maxwell asked, what about our concepts of space and time?

> To determine whether there is anything in Nature corresponding to them, or whether they are mere projections of our mental machinery on the surface of external things, is absolutely necessary to appease the cravings of intelligence. Now it appears to me that when we say that space has three dimensions, we not only express the impossibility of conceiving a fourth dimension, co-ordinate with the three known ones, but assert the objective truth that points may differ in position by the independent variation of three variables. Here, therefore, we have a *real* analogy between the constitution of the intellect and that of the external world.[40]

A "real analogy" appears to obtain between the limits of our mental conceptions and the operative geometry of physical space. The younger Maxwell's Kantian conservatism with regard to the limits of geometry would alter in the following decades with the successful development of *n*-dimensional geometries and the conceptual challenge of the fourth dimension of space.[41] But at the moment his foremost concern was to express caution about the dangers of equivocation, the potential for scientistic slippage between moral and mechanical vocabularies.

> Now the analogy between reasons, causes, forces, principles, and moral rules, is glaring, but dazzling.
>
> A reason or argument is a conductor by which the mind is led from a

> proposition to a necessary consequence of that proposition. In pure logic reasons must all tend in the same direction. . . . But when we come to causes of phenomena and not reasons of truths, the conflict of causes, or rather the mutual annihilation of effects, is manifest. . . . We see, or seem to see, a plurality of causes at work, and there are some who are content with plurality.
>
> Those who are thus content with plurality delight in the use of the word force as applied to cause. Cause is a metaphysical word implying something unchangeable and always producing its effect. Force on the other hand is a scientific word, signifying something which always meets with opposition, and often with successful opposition, but yet never fails to do what it can in its own favour.[42]

Figurative processes are called forth at the limits of comprehension, staging an uncertain confrontation between the actual and the possible, the virtual and the real. Maxwell demonstrates a penetrating awareness that, while conceptual distinctions between intellectual and physical processes must be maintained, at the same time, verbal and phenomenological problematics are inextricable from the field of scientific inquiry. "Are There Real Analogies in Nature?" also shows the philosophical parameters of a mind at home in allegory. That is, Maxwell takes note of a fundamental duality of operation in the cosmos as a whole and of the necessity for mediating that gap with provisional linguistic or ideational constructions. Clearly Maxwell brought to his scientific work a vigorous hermeneutic intelligence: "That analogies appear to exist is plain in the face of things, for all parables, fables, similes, metaphors, tropes, and figures of speech are analogies, natural or revealed, artificial or concealed. The question is entirely of their reality."[43]

I think Maxwell came in the 1860s to understand something theoretical physics soon accepted and that we anachronistically term postmodern, that the distinction between figurative simulation and literal reality reaches a kind of vanishing point, especially when the conceptual act of modeling physical hypotheses turns out at times to unlock the actual structures of things. In his further studies on electromagnetism Maxwell would encounter the uncanny power of fictive simulation. We saw that in "On Faraday's Lines of Force," mathematical analogies to hydrodynamics modeled the *geometry* of electromagnetic lines of force. His next approach to this topic, "On Physical Lines of Force" (1860–61), carried the analogical method further and to extraordinary heights. Previous investigation of electromagnetic phenomena had showed that the space around electrified and magnetic bodies was charged—currents could be inducted at a distance from the energy source—and it had also been suggested that the luminiferous and electromagnetic ethers were one and the same. But definitive evidence was not forthcoming. To investigate the interrelation of electricity and magnetism and the possible unity of their operations with the action of light, Maxwell constructed a more complex mechanical model, not just of localized lines of force but of a presumed universal medium—the electromagnetic ether. Out of this improbable model would come some major theoretical advances.

Maxwell considered the ether substance a factual fiction—convinced of its existence in some form, he was skeptical about any particular description of it, including his own.[44] To nineteenth-century physics generally, it appeared to be a necessary fiction—the material medium needed for the propagation of imponderable radiation, as sound was propagated by waves of air—but it entirely eluded any stable determination of its qualities. As an absconded substance, it acted as a kind of blank space within space, a screen available for the projection of numerous theorizations. In some of its cultural or scientistic adaptations, even among active scientists, the ether carried the spiritual projections of commentators interested in maintaining the notion of a theologized or divinely charged cosmos. However, Maxwell gave his electromagnetic ether a structure that adhered to the Cartesian and Newtonian programs of mechanical explanation.[45] Within these constraints, however, as Berkson comments, "he allowed himself to imagine an impossible mechanism which might have, nevertheless, some analogies to the true one. . . . Maxwell's mechanical model of the electromagnetic field is one of the most imaginative but least credible ever put forward."[46]

The very recherché quality of the ether—its complete impalpability, the ongoing inability to grasp and describe its material structure—actually gave Maxwell a free hand to construct his model in any way that suggested itself. The constraints upon it were all imported from other branches of physics—Newton's laws in particular—which had been experimentally grounded by means of ponderable phenomena. Beyond that the ether was amenable to mechanical modeling, for conceptually considered, it was already a mechanistic construction, out of which Maxwell could fashion an ideal theoretical machine. Akin to a literary allegorist composing the figurative narration of a moral axiom or emotional process, Maxwell realized that any given analogical vehicle was dispensable, in that many others could be devised to convey the same meanings. The ether model he constructed to investigate the possibility of a mechanical explanation for electromagnetic and optical phenomena was thus a theoretical or diagrammatic allegory embodying a complex set of imponderable physical processes.

In the cross-section of the electromagnetic ether that Maxwell drew for

"On Physical Lines of Force," the hexagons represent magnetism as vortices rotating around the lines of force (fig. 12). Siegel comments that due to Thomson's work on the "Faraday effect"—the rotation of polarized light in a magnetic field—Maxwell invested the greatest truth value in the rotatory-vortex portion of his mechanism.[47] More dubiously, Maxwell granted, the circles represent electricity as spherical particles inserted as idle wheels amid the complex of vortices and enabling them all to rotate in the same direction. Magnetism and electricity were thus unified as a complex rotational/linear medium, a network of interlocking gears with mobile bearings. The energy level or intensity of the field alters the velocity of the elastic vortices, which accordingly widen or contract, in response to Faraday's dictum that "there is a tension along the lines of force and a pressure between them."[48] In static electricity or within a dielectric insulator, the electric spheres rotate in place; in a conducting medium or electrical circuit, the balls move, and their movement represents an electric current. Finally, an elastic distortion of the electric balls creates a "displacement" current that is propagated throughout the medium as a wave effect.

Astoundingly, Maxwell's mechanical ether model enabled him to draw empirically demonstrable conclusions about the unity of electromagnetism and light. In short, he unified field physics with a heuristic thought-experiment. By subjecting this fictive mechanical allegory of the electromagnetic ether to mathematical analysis, Maxwell derived the ur-form of his famous equations of the electromagnetic field. He thus set into motion Hertz's confirmation twenty-five years later of the existence of electromagnetic waves—the origin of modern wireless telecommunications technology. Moreover, by calculating his model's elasticity, in particular, the ratio between electrostatic and electromagnetic units, he derived a figure for the propagation of electromagnetic currents that was virtually identical to the speed of light.

Siegel points out that this unification was as yet still conceptually mediated by the supposed ethereal medium itself.[49] Rhetorically stated, however, Maxwell's investigation of his model exploited a scientific version of allegorical typology. In the typological mode of allegorical reading, one text signifies its relation to another text by means of formal parallels. "When one story alludes to another by resembling its form, our attention is drawn to the form itself, the *tertium quid*. All descents to the underworld in fictions delineate a collective frame of interpretation. . . . It is the formal resemblance of a fiction to its interpretation, its frame or antitype, that forges the link."[50] In analogous fashion, Maxwell's theoretical model of the electro-

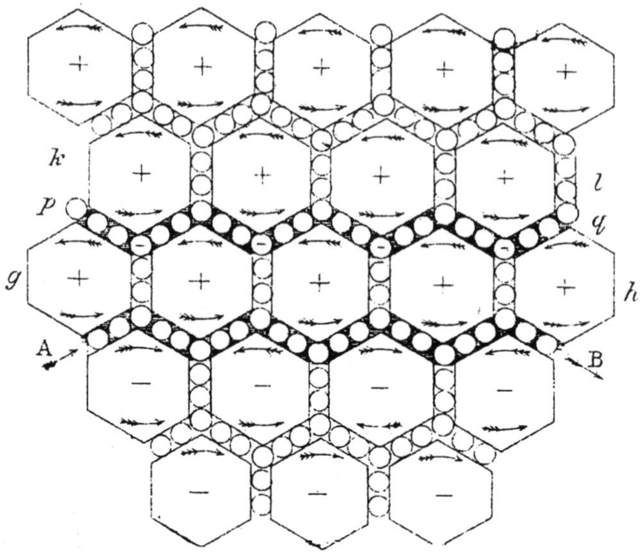

Fig. 12. The electromagnetic ether. From "On Physical Lines of Force" (1862).

magnetic medium produced a typological recognition. It generated a formal or mathematical analogy—the virtual identity of the velocities of propagation of electromagnetic currents and of light—from which followed the prediction of a valid scientific outcome—the physical unification of discrete forms of radiant energy. Maxwell himself interpreted his results like so: "The velocity of transverse undulations in our hypothetical medium, calculated from the electro-magnetic experiments of MM. Kohlrausch and Weber, agrees so exactly with the velocity of light calculated from the optical experiments of M. Fizeau, that we can scarcely avoid the inference that *light consists in the transverse undulations of the same medium which is the cause of electric and magnetic phenomena.*"[51]

But the success of the mechanical model of the ether developed in "On Physical Lines of Force" did *not* lead to Maxwell's granting it literal reality. Maxwell's prescience here, his status as Einstein's immediate forerunner, contrasts for instance to the more impetuous scientific claims of Thomson or Tyndall, many of which have long been exploded. In Maxwell's case, the figurative model remained only a productive fiction. Maxwell's accomplishment in "On Physical Lines of Force" was a tour de force of the scientific imagination, but one that still remained constrained by the very fiction that enabled its operation. Allegory, even scientific allegory, is a conceptual mode that enables significant translations among distinct levels of being or orders of phenomena, but which constrains those translations to the finitude of the particular fiction that binds the levels together.

Consequently, in his subsequent paper on electromagnetism, "A Dynamical Theory of the Electromagnetic Field" (1864), Maxwell cut the mechanical ether fiction loose, going "from the differential laws of electromagnetism, with only the electric and magnetic fields which occur there as his variables, by a direct mathematical derivation to a wave equation: no worries about whether the medium was an ideal solid, no mention of the medium at all."[52] This mathematical move beyond an embodied or mechanical ether model both anticipated Einstein's theories of relativity and confounded colleagues such as Thomson, from whom he had imbibed the method of physical analogies in the first place. Acknowledging the fictive or reified status of the ether, Maxwell simply removed it from the consideration of the literal or manifest energies that constituted the physical field. Kargon stresses that Maxwell's 1864 paper "was no longer an intermediate investigation, requiring analogies to aid the mind. It was rather a *dynamical theory,* assuming only the existence of matter in motion by which observed electromagnetic phenomena are produced. Analogies are omitted, not because Maxwell has forsaken them, but because the *method of physical analogies* was introduced for a prior stage of development of physical theory. The scaffolding could now be kicked away from the edifice."[53] What was left standing—the foundation of modern electronic technology in the abstract mathematical expressions of the electromagnetic field that developed after his death as "Maxwell's equations"—still stands.

The Demon

Among the British thermodynamicists, Peter Guthrie Tait was a prominent physical moralist and also a controversialist of the first order. His standard domestic target was the outspoken agnostic leagued with Spencer and Huxley in support of Darwinian evolution and against the influence of institutional religion in scientific matters, John Tyndall. In response to Tyndall's "Belfast Address" delivered to the British Association in the fall of 1874, which championed secular materialism against the claims of theological cosmology, in 1875 Tait and Balfour Stewart anonymously published *The Unseen Universe or Physical Speculations on a Future State,* a highly successful work of scientific popularization and Christian speculation.[54] In the midst of this detailed and innovative exposition of thermodynamic moralism, Tait and Stewart poked a grim sort of fun at the astonishment of secular Victorian scientists and intellectuals over the apparently implacable predictions of the second law. Traditional Christian allegorists, Tait and Stewart personified materialism as an evil Genius, a tempter who has seduced agnostic science into a vision of despair:

> Let us next briefly allude to the extreme school of science. Ignoring all but the visible universe, and applying the principle of Continuity to its phenomena, they were indubitably led to most important generalisations regarding the method of working of this great system. . . . So far the Genius which they had summoned up appeared to be the very principle of order. But things wore a different complexion as time went on. It was fancied that historical Christianity must disappear, and that the doctrine of immortality must follow after it. They were surrendered. But it was extremely startling when the Genius invoked, not content with what he had already devoured, insisted upon the ultimate sacrifice of the visible universe,—*then* the most extreme partisans of the school began at length to be alarmed. It was too much to be borne, that a Genius summoned up in the very name of order should turn out to be such an insatiate demon as this! Must the whole visible universe, indeed, arrive at such a state as to be totally unfit for the habitation of living beings? The individual they were content to sacrifice, perhaps even the race, but they would spare the universe. Undoubtedly, if it be possible to pity men who could so easily dispense with Christianity and immortality, they had at length got themselves into a deplorable dilemma. For, indeed, the principle they had invoked was absolutely without pity, and in the most heartless manner continued to insist upon the sacrifice of the visible universe. This, they were told, was only a huge fire, and must ultimately burn itself out. Nothing would be left but the ashes,—the dead and worthless body of the present system.
>
> No wonder, then, that these men should be startled at their conclusion, and try somehow to evade it.[55]

We will see that Tait and Stewart took evasive actions later in the *Unseen Universe,* motivated by the same cosmic prospect. And eight years earlier, Tait had helped to instigate Maxwell's profound challenge to the second law. In 1867, an exchange of letters between Tait and Maxwell brought into being another bit of scientific personification, responding directly to William Thomson's vision of dissipation, by pointing out that the predictions of the second law are probable but not certain, that the second law describes an observer's knowledge of a system rather than the system itself. Despite Maxwell's challenge, Tait retained a doomsday reading of mechanical dissipation, presumably as a stick with which to intimidate the irreligious Darwin, Spencer, Huxley, and Tyndall. But it is nicely ironic that

Maxwell's Demon, a profound forerunner of cybernetic agency, materialized in part due to Tait's own institutional and theological schemes.

Throughout 1867 Tait had been working up a series of 1864 articles on the history of thermodynamics into a short textbook, *Sketch of Thermodynamics*.[56] His text on this recently consolidated science was to intercede in the determination of scientific priorities on behalf of the Anglo-Scottish school as against the Germans. In the spring of 1867 Tait had sent draft versions of the book to Clausius and Hermann von Helmholtz, soliciting their comments on his historical constructions, which were heavily slanted toward his compatriots James Joule and William Thomson. On December 6, 1867, Tait ran a similar request past Maxwell.[57] With Maxwell he relinquished the posture of professional scientific authority he assumed in his ongoing public controversies with materialist infidels and vulgar popularizers such as Huxley and Tyndall:

> Are you sufficiently up with the history of Thermodynamics to critically examine & put right a little treatise I am about to print—and will you kindly apply your critical power to it? You would greatly oblige me by doing so, as Clausius & others have cut up very rough about bits referring to them. I don't pretend to know the subject thoroughly and would be glad of your help.[58]

Maxwell replied five days later. He declined to enter Tait's priority crusade: "I do not know in a controversial manner the history of thermodynamics, that is, I could make no assertions about the priority of authors without referring to their actual works."[59] Rather, he raised the level of the conversation by placing into doubt the absolute nature of the entropy law: "To pick a hole—say in the 2nd law of θ^{cs} [thermodynamics] that if two things are <at the same temperature> in contact the hotter cannot take heat from the colder without external agency."[60] The implied reference here is the work of Clausius, the particular focus of Tait's animosity among the Germans, and a particular critic of Maxwell's published work. However, Maxwell was more interested in testing the second law against his own version of the kinetic theory of gases and measuring the heuristic power of the statistical method he has been elaborating in that context. Using the letter to Tait as a convenient impromptu forum, Maxwell now proceeded to construct a mechanical model. The demon first materialized as the intelligence posited within, or alongside, this very simple piece of hypothetical mechanism: "Now let *A* & *B* be two vessels divided by a diaphragm and let them contain elastic molecules in a state of agitation." After further specifying the experimental apparatus, Maxwell gave his invention the daemonic turn—positing an agency in excess of the literal physical occasion—that marks it as an allegory of thermodynamics: "Now conceive a finite being who knows the paths and velocities of all the molecules by simple inspection but who can do no work except open and close a hole in the diaphragm by means of a slide without mass."[61]

In order to write off the second law, Maxwell changed the point of observation. To visualize the statistical nature of thermodynamic entropy, he imagined a being that banks on the randomized distribution of heat and cold at the molecular level entering into and sorting a closed system of energic matter. This ur-version of the demon scenario posits a partitioned vessel of gas, whose adjacent chambers *A* and *B* possess equal numbers of molecules but different temperatures. It then deploys a "neat fingered being"[62] to violate the second law by allowing heat—molecules of relatively higher kinetic energy—to pass from the cooler chamber, *B*, to the warmer, *A*, "without external agency." Under normal circumstances, such outer agency would consist of the mechanical application of an external heat source to an otherwise closed thermal system, an additional expenditure of energy in compliance with the second law.

The demon represents the imaginable possibility of an *internal* agency entering the system at the microlevel—where the averaged motions of molar masses give way to the individual vagaries of single molecules. Such daemonic knowledge allows a molecular sorting that increases the temperature differential between the chambers, counter to the entropic drift toward equilibrium. Given the convertibility of falling heat into work as dictated by the first law of thermodynamics, one would thus get mechanical work for nothing—in William Paulson's phrase, energy without expenditure.[63] "If the heat is the motion of finite portions of matter and if we can apply tools to such portions of matter so as to deal with them separately, then we can take advantage of the different motion of different portions to restore a uniform hot system to unequal temperatures or to motions of large masses."[64] It is all, in Maxwell's seminal demon scenario, a question of cleverness combined with neatness, that is, the application of an intelligence, by definition weightless and frictionless, at the proper infinitesimal scale. "Only we can't," he concludes, "not being clever enough."

But the resilience of Maxwell's Demon rests not on its technical possibility or impossibility, but on its power to operate in the interstices between physical theories and cultural desires. Maxwell's "very observant and neat fingered being" immediately departed its scene of origin and began its cultural extension as an allegory of thermodynamics, when Tait forwarded the manuscript of this letter to their mutual colleague, William Thomson. Thomson was a fateful recipient of Maxwell's heuristic fiction: fifteen years earlier he had first articulated the catastrophic notion of the cosmic consequences of the second law in his paper on the dissipation of mechanical energy. He was immediately bemused and intrigued by Maxwell's creation, as evidenced by the annotation he left on the manuscript before returning it to Tait. Opposite the end of Maxwell's remarks on the neat-fingered being, Thomson scribbled: "Very good. Another way is to reverse the motion of every particle of the universe, and preside over the unstable motion thus produced."[65]

Over the next decade Thomson took up Maxwell's fiction in a series of papers.[66] Particularly important is Thomson's 1874 paper "The Kinetic Theory of the Dissipation of Energy," which began by developing the dynamical idea he had inscribed on Maxwell's manuscript: "In abstract dynamics the instantaneous reversal of the motion of every moving particle in a system causes the system to move backwards. . . . That is to say, in mathematical language, any solution remains a solution when t is changed into $-t$."[67] Thomson elaborated this classical notion of temporal reversibility precisely by recruiting, as he now writes, "an army of Maxwell's 'intelligent demons.'"[68] Thomson tried to discipline the Demon for classical proprieties, while indelibly marking his protégé's fictive sorting-agent as an agent of physical metamorphosis.[69]

Allegories are situated between temporally divergent discourses—the clash of times is the mark of time. Just as the daemon mediates between the human and its other, Maxwell's Demon is a liminal figure stationed at a point of division within a whole system, and operating on the threshold between the conceptual spaces of reversible and irreversible time. The demon is the primal resident of a heuristic fold or implosion in classical physics, created by Maxwell's own desire to test the limits of thermodynamic laws. It marks the moment when scientific conversation first seized strong conceptual purchase on the principle of mechanical dissipation, mythopoetically, by means of a daemonic operator or allegorical model.

In its original role Maxwell's Demon represented a perfect thermal engine, frictionlessly sorting random heat energies to re-create a usable differential. Tait and Stewart's *Unseen Universe* labored both to accommodate the demon and to evade the threat it posed to their particular moralization of thermodynamics. This episode offers an example of dogmatic allegory defending against the daemonic potential of a strong revisionary trope.

The demon is first introduced into the *Unseen Universe* to underscore what Tait and Stewart retroactively presented as the caution with which Thomson had originally set forth his admonitions regarding the inevitability of universal dissipation given "man as at present constituted." In their ongoing revisionist history of thermodynamics, the authors stated: "It was Sir W. Thomson who (in 1851) first correctly adapted Carnot's magnificently original methods to the true theory of heat; and it is especially noteworthy to remark how, even at that early time, he saw the full danger of attempting to lay down any thing too definite on the subject."[70] But this danger was effectively grasped only after Maxwell conceived the being that "picked a hole" in Thomson's and Tait's previous absolutisms. Now the demon was given its due. Or more precisely, it was rewritten along Thomsonian lines—hence the *army* of demons rather than Maxwell's individual agent—and so co-opted into the larger apparatus of *The Unseen Universe.*[71] The authors noted that the velocities of particles in a vessel of gas always tend to equilibrium. However, "by enlisting in our service conceivable finite beings (imagined by Clerk-Maxwell, and called demons by Thomson), it would be possible materially to alter this state of things, even although these beings should do absolutely no work":

> For suppose a firm partition, full of little doors (themselves without mass), to be placed so as to divide the vessel into two, and set a demon at each door, with instructions to open it for an instant whenever he sees he can thereby let a quick-moving particle escape from the first compartment to the second, or a slow-moving particle from the second into the first. Then, *because the tendency is not to a uniform distribution of velocity* among the particles, but to a distribution which involves quicker and slower in certain proportions, we may imagine this process to be carried on long enough to make a considerable difference in the average velocities of the particles in the two compartments, *i.e.*, a greater pressure per square inch on the walls of the second compartment than of the first; and thus, if the partition wall were moveable, a certain amount of work might be obtained by allowing it to move. Thus a group of particles originally incapable, without external assistance, of doing work, may be rendered capable of doing work by mere *guidance* applied by finite intelligence.[72]

It is symptomatic of the hierarchical turn in Tait's thinking that the "guidance" provided in this version of the demon-scenario derives not from the demons at the doors but from the implied superdemon giving his microscopic "army" their "orders." This scene of negentropic ordering is not "self-organizing" but rather the top-down outcome of a command structure. The effect of this scheme is precisely to contain Maxwell's Demon once more within the higher determinism of a proper theologico-scientific hierarchy. On the one hand, "we see that what the demons . . . have guided the gas to do, is to transfer heat from a colder to a hotter portion of the gas,"[73] a clear violation of the second law. But on the other hand, regrettably, "the only reason why this does not occur without the assistance of demons (at least to an extent, or for a length of time, sufficient to produce a sensible effect) lies in the enormous number of particles per cubic inch in even the most rarefied gas. Hence, *solely because of the excessive numbers and minuteness of the particles of matter,* the one chance of escape from Carnot's proposition [the ur-form of the second law] is denied us" (90). For the authors of the *Unseen Universe,* the dilemma was not, as Maxwell had put it eight years earlier, that "we" are not "clever enough" to perform the necessary manipulations to escape from the rule of entropy, it was rather that the fallen material constitution of the world is bound to foil any attempt we might make. This demotion of humanity's daemonic potential to wrest new life out of base matter set the stage for a counterproposal—the existence of an "unseen universe" absolved from the mechanical destiny of the physical world.

In the full development of its argument, the *Unseen Universe* extorts from the Victorian discourse of thermodynamics an allegorical defense against that discourse's own daemonic swerves away from theological rectitude. With the demon out of the way, Tait and Stewart were free to ratify their own thermodynamic prediction of final equilibrium and universal heat-death, to embrace Thomson's vision of material dissipation, and to throw it squarely in the face of the materialists. The authors had previously asserted that there was "no escape" from their reasoning: "the visible universe must, *certainly in transformable energy, and probably in matter,* come to an end. We cannot escape from this conclusion."[74] With Maxwell's Demon impotent to interfere further with their preferred moralization of physics, the authors of the *Unseen Universe* now paint an altogether lurid picture of the material wearing down of the universe, a kind of baroque funereal displaying of the death throes and corpse of the visible world:

> The tendency of heat is towards equalisation; heat is *par excellence* the communist of our universe, and it will no doubt ultimately bring the system to an end. . . . While the sun thus supplies us with energy he is himself getting colder, and must ultimately, by means of radiation into space, part with the life-sustaining power which he at present possesses. Besides the cooling of the sun we must also suppose that owing to something analogous to ethereal friction the earth and the other planets of our system will be drawn spirally nearer and nearer to the sun, and will at length be engulfed in his mass. . . . He will be extinguished until, after long but not immeasurable ages, by

> means of the same ethereal friction his black mass is brought into contact with that of his nearest neighbour. . . . There will be (at immense intervals of time) mighty catastrophes due to the crashing together of defunct suns. . . . And so on, growing in grandeur but diminishing in number till the exhaustion of energy is complete, and after that eternal rest, so far at least as visible motion is concerned. (90–92)

The *Unseen Universe* creates a modal allegorical cosmos, a bifurcated world that equivocates at all points between physical and metaphysical realms. It is also coded with a high Victorian anxiety that scientific materialism operates hand in hand with political "materialism," thermal "equilibration" with the egalitarian dismantling of traditional social differentials. Displacing their ideological motives onto thermodynamic allegories, Stewart and Tait were content to give over the material cosmos to its revolutionary assassin, the "communist" of heat death. The *Unseen Universe* draws out the epiphany of materialistic destiny that envisions a "black mass" at the bitter end of time in order to urge its own energic reinscription of an "unseen universe" that recovers at another, metaphysical level all that was squandered in the "seen" or material world. All the *invisible* manifestations of energy—radiation, electricity, and heat—are gathered into the infinitude of the divine. Whereas Maxwell's original demon was a creature of speculative desire invoking the uncanny transformative capacities of figurative agency, the *Unseen Universe* marshaled the figure for its arguments in favor of the totalistic hierarchies of a theological allegory. These divergent applications of the demon fiction show how the allegories of Victorian thermodynamics were split at their inception between the theological moralization of entropy as material ruin and the daemonic manipulation of matter and energy in anticipation of the leap of postclassical physics beyond mechanical determinism.

Chapter 5
Dark Stars

Charles Howard Hinton's forgotten fable of 1884, *The Persian King,* presents an allegory of the fourth dimension of space in terms of the physical dilemma of entropic degradation. Hinton's own elucidation of the energic moral of his allegory, a parable of the energy bonds that unite an unseen sovereign to his subjects, ventures a remarkably prescient anticipation of the contemporary revaluation of thermodynamic entropy as a potentially creative rather than inevitably destructive physical trend. H. G. Wells's famous novella of 1895, *The Time Machine,* probably borrowed from Hinton some of its most significant conceptual texture—its evocation of the fourth dimension of space in the larger context of a temporal allegory of evolutionary degeneration and physical dissipation. It may also have derived its terminal vision of solar exhaustion from Flammarion's *La Fin du Monde* of 1893, which text laid out a series of scientistic doomsday scenarios. However, despite the Time Traveller's temporal mobility, Wells eliminated from his tale any line of escape from entropic finality. In contrast, Flammarion joined Hinton in the attempt to extract from classical thermodynamic mechanisms some form of counterentropic possibility. But whereas Hinton finessed the second law by reversing its moral polarity, Flammarion reverted to one of the earliest recuperative scenarios of classical thermodynamics, in which the first law's promise of ultimate conservation is pitted against the second law's specter of inexorable dissipation. At the end of Flammarion's narrative, in the great hereafter "après la fin du monde terrestre," the dark stars of thermodynamic entropy are rekindled in a cosmic scenario of perpetual destruction and recreation.

Thermodynamic Dimensions in *The Persian King*

In the 1880s British mathematician Charles Howard Hinton developed a seminal version of hyperspace philosophy, gathering mathematical and physical scientism together with Platonic and Kantian idealism. Hinton's essays and fables were first collected in 1886 under a title H. G. Wells co-opted a decade later, *Scientific Romances.* Over the next twenty years Hinton brought out four more works centered on the concept of a spatial fourth dimension: *A New Era of Thought* (1888), a second series of *Scientific Romances* (1902), *The Fourth Dimension* (1904), and *An Episode of Flatland* published the year of his death in 1907.

Expounding the fourth dimension through didactic parables as well as discursive presentations, Hinton's writings are a literary curiosity as well as a scientific oddity. There is an ethical edginess to his discourse, a tendency to break out of abstract geometric and physical matters into visionary lyricism and the cosmological sublime. Eliciting hypothetical dimensions by supposing away the finality of three-dimensional or Euclidean space had been a standard *n*-dimensional exercise. But for Hinton the fourth dimension of space was more than an ideal geometric construction—it was a matrix of millennial expectations. The quest to cognize higher space, Hinton believed, would call forth and establish a higher form of consciousness. Thus the individual pursuit of the fourth dimension was a moral discipline with collective evolutionary consequences.[1]

In his parables of alternative dimensions, Hinton used imaginary beings as pedagogical and heuristic tools. In this he followed numerous earlier mathematicians. Whether he was also following Reverend Edwin A. Abbott's *Flatland: A Romance of Many Dimensions* (1884) is unclear, but seems unlikely. Hinton and Abbott probably derived the immediate tableaux of their narratives from common sources in previous mathematical speculation. But the inspiration for *Flatland*'s climactic episode—in which the two-dimensional protagonist A. Square receives a startling annunciation from the higher being A. Sphere, who then lifts him into the world of three dimensions—probably came from Hinton's "What is the Fourth Dimension?" published earlier the same year.[2] Abbott may well have been satirizing Hinton's earnest dimensional millennialism, for A. Sphere descends like an angelic messenger of divine intelligence to seek in A. Square, "as being a man of sense and an accomplished mathematician—a fit apostle for the Gospel of the Three Dimensions, which I am allowed to preach only once in a thousand years."[3]

Relative to the near-obscurity of Hinton's oeuvre, *Flatland*'s quaint mathematized send-up of Victorian social and sexual mores has enjoyed a modest ongoing popularity.[4] Next to this polished and enigmatic tale by a veteran pedagogue, Hinton's rather lurching and tendentious geometric romances are often amateurish. Still, Abbott's satire is culturally tame in comparison with Hinton's uncanny leaps and dives. Moreover, Abbott had no scientific pretensions, whereas Hinton's millennial motives were always tempered by his concern for scientific plausibility, his focus on the "physical facts." In introducing his 1885 text "A Plane World," Hinton acknowledged its similarities to "that ingenious work, 'Flatland,'" but noted that Abbott's fantasy is unconcerned with the "physical conditions of life on the plane. . . . He has used them as a setting wherein to place his satire and his lessons. But we wish, in the first place, to know the physical facts."[5]

In chapter 7 we will return to Hinton's hyperspace philosophy in relation to the physical and metaphysical discourses of the luminiferous ether in Tyndall, Maxwell, Stewart, and Tait. Here we will treat Hinton's fourth dimension of space as a creative response to the cultural stresses of classical thermodynamics. In the first series of *Scientific Romances,* his long parable *The Persian King* approaches the fourth dimension of space through a virtual allegory of thermodynamics. From the standpoint of Victorian thermodynamics, Hinton's fourth dimension is a counterentropic realm discovered beyond the rule of irreversible time. This perfectly equivocal material construction dematerializes whatever it touches: a hyperbeing would have the ability to annihilate or create any object in the three-dimensional world, simply by carrying that object into the fourth dimension or placing it down once again in the third. Hinton sought scientific sanction for his system by asserting the continuity of natural laws in the fourth dimension: but when creatures from a higher dimension impinge upon lower worlds, to the lower beings natural laws *appear to be* abrogated.

The fourth dimension came to Hinton as a maternal matrix—a fecund yet invisible zone of ontological nurture—but in *The Persian King,* the hyperbeings that emerge from it are marked with idealizations of the father.[6] "If we think of a man as existing in four dimensions," Hinton wrote, "it is hard to prevent ourselves from conceiving him as prolonged in an already known dimension. The image we form resembles somewhat those solemn Egyptian statues which in front represent well enough some dignified sitting figure, but which are immersed to their ears in a smooth mass of stone which fits their contour exactly."[7] One is reminded of the hieratic massy archaism of vorticist sculpture—the monumental fixity of this image conveys a sense of vast, impersonal power—but Hinton's accent, distinctly Victorian, is on the domestic authority of the bourgeois parent. We dwellers in three dimensions can see all the points of a two-dimensional object simultaneously; similarly, in relation to a hyperspace being, our existence would also be exposed at all points to a higher-dimensional gaze. Just as a hyperbeing possesses a panoptic view of three-dimensionality, the Law of the Father institutes in the superego an intrapsychic agency that "transcends" the (male) ego while peering, as from another dimension, into its innermost secrets. The thermodynamic allegory of *The Persian King* carries with it a familial romance through which Hinton negotiates with, and to some extent subverts, the patriarchal inscription already evident in the Victorian codification of thermodynamic axioms *as* authoritative laws.

The Persian King is perhaps Hinton's most ambitious single text. Its hundred pages draw into a baggy unity Hinton's dual and often clashing modes of expository declamation and allegorical moralism, and foreground Hinton's investments in physics as well as mathematics. Moreover, it implicitly links his version of the fourth dimension to the scientific milieu of Faraday, Kelvin, and Maxwell, long before he did so explicitly in his 1902 article "The Recognition of the Fourth Dimension."[8] *The Persian King* represents Hinton's inaugural effort to attach hyperspace to the world of late-classical physics as a plausible extrapolation from the theory of the ether. When he addressed physics in his other, explicit treatments of dimensionality, Hinton usually focused on the electromagnetic/luminiferous ether of Maxwell's field theory. *The Persian King* is unique in demonstrating the late-Victorian pathways that also connect the fourth dimension and the ether hypothesis to the discourse of general thermodynamics. Part 1 presents the extended vehicle of the allegory, a fable of energy in terms of the rise and fall of a mysterious isolated civilization; in part 2 Hinton explicates his fable with a revisionary reading of Victorian physics. In the following synopsis I will occasionally collate the two parts.

While hunting one day the Persian king comes upon a cloistered valley with no entrance but a rock bridge over a chasm. As the king crosses, the bridge collapses, imprisoning him in an unknown land. That night he encounters "Demiourgos"—an aged man with a rustic pipe, who creates new worlds with his mystic melodies. Demiourgos gives the valley and its newly created inhabitants to the king as raw materials on which to work cultural improvements. For at the moment, the beings of the valley are perfectly inert. The old man explains that "the beings as I can make them, they follow pleasure and avoid pain. And if the pleasure and the pain are equal they do not move one way or the other."[9]

In the manner of his hyperspace narrations, Hinton's allegory in *The Persian King* establishes an ontological hierarchy of higher and lower beings—the old man, the king, and the inhabitants of the valley—with the king mediating between higher (four-dimensional) and lower (three-dimensional) realms. The reader is eventually informed that the subjective economy of pleasure and pain that determines the activities of the valley-dwellers is meant to convey the objective exchange of physical energies, the interplay of conservation and dissipation of energies under thermodynamic laws. In order to make the beings perform work, the old man tells the king, he must assume a bit of their pain, and so produce the surplus of pleasure that tilts them into activity. In Hinton's allegory of energy, to bear pain means to absorb dissipated energies.

From Demiourgos the king learns how to set his new subjects into motion, guiding them toward various actions by removing their pains, while minimizing the amount of pain he assumes from them. Turning his staff into a pendulum with which to demonstrate reciprocal dynamics, Demiourgos initiates the king into the mechanisms of energy: "For regard my staff as it begins to swing. It is not I that make the movement that is imparted to it; that movement lay stored up in my arm, and when I struck the staff with my arm it was as if I had let another staff fall which in its falling gave up its movement to the one I held in my hand."[10] This is the first law of thermodynamics: the totality of energy remains constant while passing along in various forms from one manifestation to another. When the king asks about the fate of energy that has passed out of our grasp, he is told: "It goes to the finer particles of the air, and passes on and on. There is an endless chain. It is as if there were numberless staffs, larger and smaller, and when one falls it either raises itself or passes on its rising to another or to others. There is an endless chain of movement to and fro, and as one ceases another comes. But, O king, I wish to take thee behind this long chain" (1:42).

The old man's reply anticipates what Hinton will call "the central question—the significance of the passing away of energy"[11]—what we would term the moralization of the second law of thermodynamics. In Hinton's scientific cosmos, to go behind the long chain of energic transformations is to emerge on the far side of the ether. In "What is the Fourth Dimension?" Hinton had suggested that the fourth dimension may be thought to impinge upon our world in its "ultimate" particles, to which the processes of dissipation finally transport the kinetic energies of matter. In *The Persian King*, Hinton's "ultimate medium . . . the last and ultimate substance" is the unstated fourth dimension: "in point of speed of transmission the properties of this ultimate medium must be infinitely beyond those of luminiferous ether. To this ultimate medium all movements at any distance from each other must be almost equally present at every part" (1:118–19).

In the 1880s, the wave theory of radiant propagation in the medium of the luminiferous ether and Maxwell's electromagnetic field theory were the most advanced positions in energy physics. It was a short step from those theories to the conception of the fourth dimension as an instantaneous connective space beyond material constraints. Situating his implied discourse of dimensionality within the matrix of field physics from which modern space-time would later emerge, Hinton inscribed his tale with a hovering emblem of the electromagnetic field, when Demiourgos crowns the king with a circle of rays by which to bind up his realm: "'Take that,' the old man cried. 'The rays go forth unto everything in the valley. They pass through everything unto everything. Through them thou canst touch whatsoever thou wilt.'"[12] But the rays are also the puppet-strings with which the king will perfect his energic manipulations and lead the valley-dwellers from aboriginal inertia to active civilization. The relation of the valley subjects to the king embodies the susceptibility to manipulation, the puppetlike dependency, that would exist in the relation of three-dimensional beings to a hyperbeing. The humanoid valley-dwellers, reduced to the level of complex automatons, are telling emblems of the mechanistic vision of humanity that haunts this fable of higher ideality.

The king now proceeds to attach the rays to his subjects and, by compassion, to set the valley into motion, suffering the differential of pain in all activities, while maintaining himself unrevealed. So that the inhabitants can move beyond the most elementary routines of existence, they are endowed with a mechanical aptitude for figurative processes: "For they had a sense of analogy, and observing some activity which the king had led them through on a small scale, and in which they had found a balance of pleasure, they were ready to try a similar one on a larger scale."[13] Allegory theory posits that daemonic creatures are "possessed by" and thus bear the insignia of the conceptions they are deployed to represent.[14] Here Hinton projects upon his mechanistic beings the very mechanism of their literary production—the same knack for analogy that has brought them into allegorical being in the first place.

The cosmogonic and archaic phases of Hinton's allegory are completed as the valley develops into a facsimile of a traditional Western society much like preindustrial England. Now the narrator begins to circle around a few issues of religion and science—the emergence of certain prophets with various beliefs about the unseen king, and the rationalization of the valley-beings' "typical routine." Hinton repeatedly suspends his narrative to work through an analysis of the mechanism of this "fundamental activity":

> Hence the three points which were characteristic of the activity of the beings in the valley are obvious enough.
>
> 1. There is as fundamental type a routine AB, AB, AB, the sensation involved in which goes on diminishing.
>
> 2. There are routines CD, CD, &c., connected with AB, AB, in which the sensation which disappears in the routine AB, AB seems to reappear.

> 3. In the action AB itself there is a disappearance of sensation. The sensation connected with A is 1000, that connected with B is 998. Thus 2 of sensation seems to have disappeared. This 2 of sensation is of course the pain which the king bore, and which was the means whereby the creature was induced to go through the action at all. But looked at from the point of view of sensation, it seems like a diminution of amount.[15]

As encoded in this affective idiom, Hinton is again addressing the physics of dissipation in terms of a continuous series of energy conversions from the visible world of mundane activities to the molecular level of ultimate particles. Entropy is presented here as "diminution of sensation," the increment lost in the repetitive oscillations—"AB, AB, AB"—of pleasure and pain, that is, kinetic and potential energy, as in the frictional running-down of a swinging pendulum. When the scientists of the valley discover for themselves the laws of thermodynamics, Hinton reinscribes the affective symbolism in a physical idiom, as the conservation and loss of *vis viva,* "living force": "They found, as nearly as they could measure, the routines which sprang up as the routine A B died away were equal in sensation to the loss of the routine AB, AB. And from this they concluded that the amount of sensation or feeling was constant. They called it living force. . . . But after a time, with more delicate measurements and more patient thought, they found that some of the sensation was still unaccounted for."[16]

The scientists of the valley reach the dilemma posed in 1852 by William Thomson's "On a Universal Tendency in Nature to the Dissipation of Mechanical Energy"—the impasse of a mechanistic physics, nurtured on Newtonian dynamics, when confronted with ultimate energic waste due to the irrational and inexorable friction of mundane physical processes. The fourth dimension of space has come, to Hinton and to late-Victorian culture, in answer to this thermodynamic anxiety. The tale implicitly suggests that the existence of a fourth dimension can resolve the mechanistic dilemma of devolutionary waste. As the old man had done for the king, the narrator of *The Persian King* now offers to take us "behind" the chain of mechanistic processes, to discover that what at one level appears to be mere dissipation is on another level a form of cosmic tolerance for all becoming. The problem lies not with the cosmos but with the provincial, constricted perspective of the valley's thinkers, due to which the solution to the threat of entropy—the higher reality of the hidden king—is unavailable to the culture "as at present constituted."[17] A prophet must come forth to break through the limitations of a circumscribed rationality.

The narrative of this imaginary civilization turns now to the tale of a university student expelled for expressing heretical doubts about the official science of the valley. The unnamed student is exiled to a hamlet on the perimeter of his world with "a peaceable race of savages, engaged in agriculture,"[18] where, after meditating on folk traditions in the context of the science he has learned, he has a revelation. In this theodicy of entropy, "the presence of the pain in the valley would prove that this power took only some of the pain and not all. . . . The being who, bearing pain, gave existence to the inhabitant, used economy in his actions—he chose to effect his objects with the least possible expenditure of means."[19] Such a vision of cosmic functioning echoes, at a speculative remove, other systems of extended energetics proliferating in the culture of Hinton's time—unitary systems that formalized the physical, biological, social, and industrial economies of energies.[20]

Although the omniscient king knows that the student has intellectually grasped his higher existence, he still withholds himself. In the familial complex now breaking rudely into the text, Demiourgos represents a transcendental creative power; the Persian king as a master manipulator of that power is a projection of the father; the troubled but prophetic student is the author's own self-projection. And at the climax of the psychoscientific allegory, the noncommunication of the king-father thwarts the mission of the student-son. The student turns back toward the central city, acquiring a disciple along the way. But the disciple is unfaithful, and the student is put to death for subverting the laws of pleasure and pain.

So the allegory proper ends with a cautionary moral—ostensibly to the valley-dwellers, but implicitly to the king-father—about the cost of spurning the son-visionaries of the valley. The martyred son possessed a secret that the civilization of the valley was unable to grasp. Its failure to acknowledge or credit the student's vision of a compassionate god—that is, to realize the resources of the fourth dimension—denies to the valley a form of salvation from the fate of passing away into mechanical apathy. For soon after the killing of the prophet, the civilization of the valley spirals into an entropic free-fall. In this final parable of cultural fatigue, "the natural spring of life . . . seemed tending to fail in every one,"[21] until at last the universal lassitude infects the king as well. Moving with the instantaneous leaps of the four-dimensional, Demiourgos returns for the exhausted ruler:

> Crossing by an unknown way he came and stood by the king's side. After a while the two moved on together, and by a secret path passed away from the valley—whither I know not.
>
> As soon as the king had departed from the valley the beings in it began to sink into the same state of apathy as those were whom he had first found there. Those who sank first were the ones in whose lives the stress of labour

> or thought was the most intense, for they first felt the loss of that bearing of pain by one beyond themselves which gave them a difference of pleasure. And slowly as the accumulated enjoyment was exhausted, a chill death in life crept over the land. 'Tis useless to ask after the fate of any one of those that were there, for each was involved in the same calamity that overwhelmed all. Every hand forgot its cunning. The busy hum of life in the streets was hushed. In the country the slowly moving forms gradually sank to rest. At every spot was such unbroken quiet as might have been had all the inhabitants gone to some great festival. But there was no return of life. No watchful eye, no ready hand was there to stay the slight but constant inroads of ruin and decay. The roads became choked with grass, the earth encroached on the buildings, till in the slow consuming course of time all was buried—houses, fields, and cities vanished, till at length no trace was left of aught that had been there. (1:100–101)

The manifest apocalypse of Hinton's allegory of thermodynamics starkly dramatizes the Victorian moralization of entropy as God's withdrawal from the material world. However, the broader cultural significance of Hinton's literary sublimations is that the full text of *The Persian King* also subverts this typical moralization of physical dissipation, by revising entropy as "the ultimate permission." The Persian king and the valley are punished precisely for their failure to grasp a message of salvation. Although unlistened to, the student had nonetheless discovered a higher law. Hinton's intimation here is in line with James Clerk Maxwell's crucial if debatable observation that entropy is not a substance like matter or a dynamism like energy, but an epistemological effect, a product of the limitations of human perception—for instance, our inability to manipulate matter at the molecular level. Intriguingly anticipating more recent views, Hinton's allegory posits entropy—"the pain of the king"—as a partner of creation rather than its destroyer. The king who converts this pain into life—the unseen master who arrives from and departs to another world—personifies the master subtext of this narrative—the four-dimensional matrix that receives the ultimate oscillations of dissipated energies, and out of which infinite universes can be reconceived. The king can bestow the gift of energy because he takes the pain of entropy upon himself: entropy is revalued not as mere dissipation but as a creative motive for collective redemption through the virtual construction of the fourth dimension.

Hinton had to confront thermodynamic laws in order to make conceptual room for the fourth dimension, a physical hypothesis allowing for material events that violate normal thermodynamic constraints. In *The Persian King* he gave the second law the slip by yielding to it so perfectly that it turned into the enabling condition of all motion and thus all energy: "the passing of the motion of masses into the form of heat is the ultimate permission."[22] This unusual championing of dissipative processes—an appreciation for, rather than denigration of, friction and resistance—is the truly predictive portion of Hinton's text. An axiom of chaos theory is that without friction, self-organization cannot commence.[23]

> This passing of energy into the form of heat must not be regarded as a side circumstance, as less essential to the laws of nature than that law which we call the conservation of energy. It is at the same time the end of every motion, and that which makes every motion possible. The passing of energy into the form of heat takes place in that which we call friction, and in all those modes in which any movement is bought to a standstill. But so far from these being simply "hindrances" to motion, it is through them that we learn that which makes motion possible. . . . Let us find in that mode whereby all motion comes to an end the originating cause also whereby all motion comes to be.[24]

Hinton submits the world to the randomizing processes of entropy, but in a final idealistic turn beyond the "physical facts," reverses its metaphysical valence with the postulate that "that which in material terms we represent as an infinite series [in this context, the dissipation of energy] is a will—a will in contact with all existence, as shown by the properties it had when we conceived it as an ultimate medium."[25] This maneuver plots the location in Hinton's system for a four-dimensional agency that restores true causation to the cosmos. In the end, the image of hyperagency that grounds Hinton's allegory of thermodynamics is set forth as a human spiritual goal, and the powers responsible for the initiation and direction of cosmic energies are invested in a myth of personality. The ultimate communication in Hinton's system is between higher dimensionality and transcendental personality. The aim of the quest in this scientific romance is to transcend the status of puppet in regard to "material sequences" by rising to a position of self-determination. Higher space contains this absolute self, a telos that transforms the Kantian epistemology from which Hinton began.[26] The self-exegesis of *The Persian King* ends by conflating Hegelian sublation with the Schopenhauerian transumption of the material universe by an absolute will:

> Does this will not exist in those who are true personalities, and not mere pleasure-led creatures?—have they not some of this power, the power of accepting, suffering, of determining absolutely what shall be?—a creative power which, given to each who possesses it, makes him a true personality, distinct, and not to be merged in any other—a power which determines the chain of mechanical actions, of material sequences—which creates it in the very same way in which it seems to be coming to an end—by that which, represented in material terms, is the absorption of energy into an ultimate medium; which,

> represented in terms of sensation, is suffering; but which in itself is absolute being, though only to be known by us as a negation of negations.[27]

This final twist in Hinton's exposition outlines a dialectical approach to creative selfhood that anticipates modernist aesthetic manifestoes calling for an art that combines energic and formal efficiency with visionary self-assertion. In Hinton's writings, mathematical physics morphs into incipient science fiction. He was essentially a scientific romancer on the order of Verne or Wells, as yet too enmeshed in scientific and philosophical agendas to cut his texts loose as pure fictions. The strains that play through his writings are the labor pains of science fiction as that genre struggled to raise itself to self-conscious status from out of the Victorian matrix of scientistic speculations. An important precursor of the allegorical paraspaces of science fiction and the dimensional warps of virtual reality, the fourth dimension of space has been revived in recent science.[28] A generation before the "four-dimensional" continuum of relativistic space-time, and a century before computer screens, it was cyberspace *avant la lettre*—a potentially collective thought-construction inhabited by vast structures, virtual continents beckoning to be explored. Tracing the logic of late-Victorian desire in the overdeterminations of Hinton's writings, we discern some of our own paranoid fascinations with multidimensional technocultural spaces.

Allegorical Mechanics in *The Time Machine*

For many decades now *The Time Machine* has been a charged medium of cultural anachronisms. In the preface to a 1931 reissue, H. G. Wells himself already considered his youthful tale dated. Recalling the intellectual climate of the tale's gestation in the early 1890s, Wells declared, "the geologists and astronomers of that time told us dreadful lies about the 'inevitable' freezing up of the world—and of life and mankind with it. The whole game of life would be over in a million years or less. They impressed this upon us with the full weight of their authority, while now Sir James Jeans in his smiling *Universe Around Us* waves us on to millions of millions of years."[29]

The Time Machine remains a vivid cultural vehicle of specific scientific ideas from the conceptual world of 1895, but its edge as a minor yet seminal work of modern literature now rests on a proper appreciation of its modal and historical form as a virtual allegory of classical thermodynamics. Twisting together powerful strands of defunct science, its complex of superannuated scientific concepts is clustered at a discursive node that connects two obsolete formations—biological degeneration and heat death. The very images that the science of *The Time Machine* projects upon the future—multiple scenes of entropic and devolutionary ruin—are now themselves in conceptual ruin. However, the medium of allegory is geared precisely to this interdependence of the superannuated and the perennial, the obsolete and the abiding, and so provides an important vehicle for the interweaving of historical phases in science and literature.

The allegorical mode is inherently mechanical. Like any machine, it is a datable form of deliberate or de facto artifice. Allegory is mechanical not just aesthetically—as in the conceptual fixity of its traditional characters (the "stiffness" of a personification in a poem or novel, or of an emblematical figure in a woodcut or painting)—but also fundamentally in its articulation of some measured synthesis of the world, some motivated rearrangement of its real and ideal parts. Allegorical texts are cultural machines for the productive conversion of moral forces. By the same token, machines are allegorical objects, not just as commodity fetishes, but also as embedded with the social prehistories of their conception and development and the powers they mobilize upon use.

Bruno Latour writes in *Aramis; or, The Love of Technology,* "Every machine is scarified, as it were, by a library of traces and schemas."[30] Fictive machines—the visionary engines of science fictions—especially spacecraft like the *Integral* in Zamyatin's *We* and "vehicles" of daemonic flight like Wells's Time Machine, are all the more pure versions of hybrid allegorical contrivances. The Time Machine in particular is a cultural allegory of the technological prowess to conceive and construct mechanical devices for the productive conversion of physical energies. It also ensures the Time Traveller's arrival at a prophetic perspective from which to perceive the future fate of humanity and the earth as determined by the cosmological mechanism of entropy.

The Time Machine's allegorical mechanics begin with the first sentence's gesture of emblematic naming, bypassing round characterization in favor of parabolic flatness and conceptual shorthand: "The Time Traveller (for so it will be convenient to speak of him) was expounding a recondite matter to us."[31] Then a set of bourgeois characters is sketched in by professional title—the Psychologist, the Provincial Mayor, and the Medical Man—to be joined a week later in the second chapter by the Editor and the Journalist. With the exception of the reader's designated emotional proxies, Filby and Weena, all other proper names are essentially conceptual puns: "Blank" and "Dash" are (typological) forms of written or typographical spaces or (pun)ctuation marks, and "Chose" is legalese for an object of personal property. If, as Maureen Quilligan argues, "allegorical narrative unfolds as a series of punning commentaries," the chose or "thing" this story will be about is a Swiftean dash or mark of accelerated time across a blank space of fictional text.[32] The dash as a rapid movement in space also marks the narrative's central figure of daemonic flight, the Time Traveller's temporal velocity through a "space" of time.

Quilligan has drawn attention to another common device by which a text advertises its allegorical status. A "threshold image" often takes literal shape as some kind of portal or framed vista through which a visual tableau or natural landscape is transformed into a cultural text, a script to be deciphered as the narrative advances.[33] The Time Traveller's precipitate entry into the world of the future passes through such a threshold, a kind of natal portal, as he drops from his time warp and is delivered into a scene centered on an emblematic image with oblique implications. Abruptly braking his daemonic flight through the fourth dimension, he tumbles off the time machine beneath "a colossal figure . . . of white marble, in shape something like a winged sphinx."[34]

Like the winged Sphinx, allegories hover on the border between myth and science, and "mythical spaces are chimerical." In his essay "Language and Space: From Oedipus to Zola," Michel Serres locates the Sphinx at and as "a crossroads, with wings that protrude for one who no longer needs feet. The Sphinx is a bifurcation, and conversely. And the crossroads is a chimera."[35] The mythic Sphinx doubles the Time Traveller, a technoscientist who has enacted a spatiotemporal crossing-over and so created a fantastic hybrid of times, bridging an uncanny conjunction of spaces. Allegories arise from such mediating and systematizing of chaotic spaces: they create virtual topologies in which, in Serres's words, "I can no longer draw the line between narrative, myth, and science."[36] The figure of the Sphinx marks the cultural insemination of science into allegory, science's continued gestation within the matrix of literary romance. The White Sphinx's manifest enigma now poses *The Time Machine* itself as a "scientific romance" that confronts experimental questioning with allegorical and oedipal questing. The White Sphinx is just one of many modal markers, but it is the one in particular by which Wells poses the question of science *for* allegory. By his time science per se is the face of time's mysteries. Hinton's and Wells's scientific romances mark specific moments in the modern unfolding of the mystery of entropic time. The Time Traveller's quest-allegory moves precisely toward particular social myths of late-Victorian science. This sphinx is pregnant with those ideologies of Victorian thermodynamics that mystified history through scientistic claims about a foreshortened future for all of nature.[37]

When the Traveler tries to read this emblematic inscription hovering over the veiled landscape, his expectations of a stable or definitive revelation are consistently thwarted. His attempts to comprehend this world under the aegis of positive scientific understanding constantly collapse before the indecipherable text the Sphinx foretells. In the near term, what the Traveler lacks is the knowledge of the Morlocks. Mastering that subterranean subtext to the ostensible script of the landscape eventually allows him, after a fashion, to read the *economic* relations that invisibly sustain the pitiful paradise of the Eloi. Reversing the superhuman evolution of the Vril-ya in Bulwer-Lytton's *The Coming Race* through figurative devices descended from Swiftean satire, *The Time Machine*'s allegory of social Darwinism transforms Gulliver's Lilliputians and Yahoos into the devolutionary Eloi and Morlocks of 802,701. Wells's dark amalgam of late-Victorian scientisms satirizes British social inequities as well as dashes the hopes of socialist progressives that humanity's spiritual elevation to ever higher states of organization and being would follow irreversibly from the elimination of those inequities.[38] The only eternity allowed by classical thermodynamic entropy was an ultimate and permanent disorganization, a return to mere chaos.

The emblematical tenor of *The Time Machine* also derives in part from an allegory of mathematical physics, in particular, the narrative extension and connection of geometrical and thermodynamic ideas. As popularized in the 1870s by literary allegories of physics like Tait and Stewart's *The Unseen Universe* and in the 1880s by parables of mathematics like Abbott's *Flatland* and Hinton's *Scientific Romances,* the geometry of the fourth dimension of space offered a form of imaginative escape from thermodynamic confinement, an ideal retreat from the threat of imprisonment by the inexorable second law. The spatial fourth dimension made conceptually or virtually possible a "disappearance" or daemonic flight away from the entropic ruin of the material world.

The opening sequence of *The Time Machine* anticipates the Time Traveller's final disappearance by staging the vanishing of a miniature model of the actual time machine, right at the climax of a learned discussion of the fourth dimension. The fourth dimension figures here precisely as an avenue of escape, a bursting of cognitive confinements figured as a transcendence of physical constraints, as in Simon Newcomb's rehearsal of the geometry of hyperspace. Just "as the man confined in a circle can evade it by stepping over it, so the mathematician, if placed inside a [three-dimensional] sphere in four-dimensional space, would simply step over it as easily as we should over a circle drawn on the floor."[39] Classical thermodynamics determines that the escape in question for Wells is not from spatial but from temporal confinement, while classical dynamics informs the four-dimensional geometry that enables Wells's hero to invent his daemonic machine, one that converts spatial into temporal flight by converting time into space. The creation of a machine that can spatialize time secures the linear reversibility that drives of the metalepses of the narration—the Time Traveller's ability to leap from world to another, back and forth in time, as along any other, spatial dimension.

Thus the Time Traveller reproves Hinton's spatial theories for their impracticality. "Some foolish people," he asserts, "have got hold of the wrong side of that idea":

> "Some philosophical people have been asking why *three* dimensions particularly—why not another direction at right angles to the other three?—and have even tried to construct a Four Dimensional geometry. Professor Simon Newcomb was expounding this to the New York Mathematical Society only a month or so ago. You know how on a flat surface, which has only two dimensions, we can represent a figure of a three-dimensional solid, and similarly they think that by models of three dimensions they could represent one of four—if they could master the perspective of the thing. See?"[40]

In fact, Hinton had intended the spatial fourth dimension to be a literal mindbender, a realm to be explored by introspective mathematical meditation. This vortex of visionary torque, if one could really "master the perspective of the thing," would lift the speculative intellect out of physical constraints and up to a more comprehensive, metatemporal vista. Wells swerves away from the spatial fourth dimension of Hinton and Newcomb by applying speculative mathematics to an engineering project that factors the irreversible temporality of the second law into the reversible spatiality of the fourth dimension. "The geometry that they taught you at school is founded on a misconception. Any real body must have extension in *four* directions: it must have Length, Breadth, Thickness, and—Duration" (3–4). And duration bends toward entropy. But whereas Hinton's fourth dimension of space made conceptually or virtually possible a disappearance or daemonic flight away from the entropic ruin of the material world, for Wells the temporal fourth dimension is counterentropic only if one goes back to the past. Instead, the Time Traveller's adventure climaxes thirty million years into the future with a vista of terminal mechanical dissipation.

Wells's treatment of the fourth dimension as temporal rather than spatial is the master stroke of the novella. It anticipates the development from late-classical hyperspace to current models of space-time. But functionally within Wells's epochal narrative, the reversibility of his temporal fourth dimension straddles the grooves of the spatial fourth dimension.[41] Thus the "time machine" is actually an oblique form of "space vehicle." Its dashes through time are tracked in terms of the Traveler's three-dimensional spatial motions, caused by the virtual velocities of the earth's dynamics within the solar system, as the time machine speeds up and slows down within the fourth dimension of time. Passing out of his proper world by miraculous transportation and tripping through time allows this cosmic Traveler to perceive and graph the periodicities of the earth's rotations and revolutions. The temporal velocity of the time machine creates a relativistic perceptual effect, a variable oscillation of the sun.

The sun's apparent paths overhead diagram a kind of phase-space portrait of celestial motions. Fast-forwarded by temporal acceleration, the normally imperceptible motions and processions of the solar system make cosmic tracers. At the temporal velocity of a year per minute, for instance, "the sun belt swayed up and down, from solstice to solstice."[42] Captured against this variable zodiac, the Traveler is an angel of thermodynamic time, a dis- and re-appearing allegorical mediator. Much as A. Sphere had lifted A. Square out of Flatland into the realm of the third dimension, the Time Traveller passes out of his proper world by miraculous fourth-dimensional transportation, crosses over to the far future, and, eventually, brings back to his present day a dark warning about the shape of the world to come.[43] So too *The Time Machine* goes back and forth between a sociopolitical allegory of biological devolution and a thermodynamic allegory of energic dissipation.

The central episodes of biological allegory in the world of 802,701 are framed by a sequence of cosmological passages in which the mechanical exhaustion of both the solar engine and the kinetic energy of the solar system dramatizes the thermodynamics and the solar physics of Lord Kelvin, one of those who told Wells's world "dreadful lies about the 'inevitable' freezing up of the world—and of life and mankind with it." When the Time Traveller is in actual temporal motion upon his machine, the sun becomes an oscillating center, a perceptual reference point tracing these signatures of classical entropy. When first he set out on the device, the Traveler recalls, "I saw the sun hopping swiftly across the sky. . . . The jerking sun became a streak of fire, a brilliant arch, in space."[44] Although the speed of oscillation of the sun's arch is a relativistic perceptual effect of the temporal velocity of the time machine fixed upon a planet in complex motion, the apparent dynamism of this vigorous sun doing celestial calisthenics encodes the relative "morning" of the world from which the Traveler departs. He arrives in the world of the White Sphinx, however, under an afternoon sky that soon fades in "the warm glow of the setting sun" (28). After encountering the effete race of the Eloi, he reflects, "The ruddy sunset set me thinking of the sunset of mankind" (31).

At the climax of *The Time Machine,* the Traveler's dead end at Terminal Beach implies that his individual ability to move within time has given him no ultimate reprieve from the larger cosmic catastrophe. The Eloi and Morlocks are merely evolutionary premonitions of the final exhaustion of cosmic as well as vital energy, a mere thirty million years into the future. Once again the sun marks the modulation, as the Traveler leaps forward to the verge of the entropic exhaustion of the solar system. Harassed by gargantuan crustaceans, the Traveler presses on again until, "more than thirty million years hence, the huge red-hot dome of the sun had come to obscure nearly a tenth part of the darkling heavens. . . . There were fringes of ice along the sea margin, with drifting masses further out; but the main expanse of that salt ocean, all bloody under the eternal sunset, was still unfrozen."[45]

Wells's crepuscular landscape is one of the classic vistas of thermodynamic scientism—an emblem of dead energy, the ruin of nature promised by the apocalypse of ever-increasing entropy in a closed universe. Citing the thermodynamics of Thomson and Tait in their authoritative *Treatise on Natural Philosophy* of 1867, Wells's fin-de-siècle American contemporary Henry Adams lamented that "all nature's energies were slowly converting themselves into heat and vanishing into space, until, at the last, nothing would be left except a dead ocean of energy at its lowest possible level."[46] In terms of the scientistic tropes that energized the cultural field of 1895, Wells's Terminal Beach is a visionary view from the shore of this dead ocean, the vast thermal sink of dissipated energies whose sluggish waves lie inert within the coagulated ether beneath the darkening star of Victorian solar physics. And although we no longer live under the hectic and hastily dying sun of classical thermodynamics, the narrative force of *The Time Machine* as a taut classic of creative scientism is hardly less potent for the obsolescence of its scientific substance. But its brash vitality as a literary fiction demands that its documentary components be clearly grasped in their finite historical function of discursive mediation between its own era and the present. Readings that can recontextualize obsolete science fictions within the history of scientific concepts they keep in play are best positioned to remediate their critical potential for postclassical audiences.

Dark Star Crashes: "The Star" and *La Fin du Monde*

Certain allegories of thermodynamics operated on the first as well as the second law to moralize the conservation as well as the entropic dissipation of energy. To revise T. S. Eliot's modernist expression of entropic melancholy in "The Hollow Men," in some thermodynamic fables the world ends not with a whimper but with a bang. These more robust visions of destruction and ruin arose spontaneously from the imagery of competing energies and the contemplation of colliding masses. In tandem with heat death and other notions of energic fatigue, a separate register of thermodynamic catastrophe developed from speculations about the cataclysmic collision of cosmic bodies—asteroids, comets, planets, and stars. What happens when the kinetic energy resident within a physical system is suddenly dissipated in a chaos of disorderly fragments? In particular, what happens when massive bodies in extremely rapid motion hurtle into each other? If despite this sudden leap into higher entropic states, the sum of their energies of motion is conserved, where does that power go when the explosion is over? And what might these physical considerations have to do with modern hopes and fears over individual and cultural progress or survival? For some scientistic speculators, the first law of thermodynamics held out the promise of violent transformation rather than utter annihilation and thus salvaged creative hope out of entropic despair.

In *La Fin du Monde* and again in H. G. Wells's "The Star," another tale in which a comet threatens to collide with the earth, it is implied that episodes of vast destruction can release more energies than they consume, repairing cosmic fatigue with new resources of power.[47] In "The Star," a compact parable published in the Christmas 1897 number of the *Graphic,* the star that rises upon the world is not a herald of peace but a threat of destruction. The monster comet collides with Neptune, "and the heat of the concussion . . . incontinently turned two solid globes into one vast mass of incandescence."[48] Then the augmented projectile takes dead aim at Earth. Its malignant heat raises flood tides and typhoons that devastate the Pacific hemisphere, but a direct hit is averted when the moon interposes itself, takes the brunt of the impact, and deflects the star, which then dashes on "the last stage of its headlong journey downward into the sun. . . . And as the storms subsided men perceived that everywhere the days were hotter than of yore, and the sun larger, and the moon, shrunk to a third of its former size, took now fourscore days between its new and new. But of the new brotherhood that grew presently among men, of the saving of laws and books and machines . . . this story does not tell."[49]

Wells's coy short story recapitulates in miniature the shape and detail of *La Fin du Monde,* the second part of which details a "new brotherhood" in a post-cometary futurity that carries the story of human progress forward to a final demise ten million years in the future.[50] At first, however, *La Fin du*

Monde opens upon a state of global lassitude induced by the rogue comet that has emerged from deep space on a collision path with the earth. "For more than a month the business of the world had been suspended" as if under the paralysis of a death sentence: "The courts themselves had no cases; one does not murder when one expects the end of the world."[51] When the comet is just days away from impact, as the speakers at the Parisian colloquium debate the possible effects of the comet's solid and gaseous elements, the first law of thermodynamics is repeatedly invoked in the particular form of the mechanical conversion of the energy of motion into heat. The president of the Astronomical Society reviews the basic science involved:

> "The audience before me is too well informed not to know the mechanical equivalent of heat. Every body whose motion is arrested produces a quantity of heat expressed in caloric units by mv^2 divided by 8338, in which m is the mass of the body in kilograms and v its velocity in meters per second. For example, a body weighing 8338 kilograms, moving with a velocity of one meter per second, would produce, if suddenly stopped, exactly one heat unit; that is to say, the quantity of heat necessary to raise one kilogram of water one degree in temperature."[52]

A moving body is typically brought to a sudden stop by colliding with an equal body moving with an opposite velocity.[53] This detail of Victorian physics also inflects Wells's "The Star," at two points. The first is Wells's own didactic device, early in the story, for placing physical concepts before the reader: "The schoolboy . . . puzzled it out for himself—with the great white star, shining broad and bright through the frost flowers of his window. 'Centrifugal, centripetal,' he said, with his chin on his fist. 'Stop a planet in its flight, rob it of its centrifugal force, what then? Centripetal has it, and down it falls into the sun!'"[54] Here the idea of the sudden arrest of planetary motion is conceived in its inertial rather than thermodynamic aspect. The latter consideration returns at the end of the story, at its second point of intersection with the first law of thermodynamics, which is applied now to a defunct phase of solar physics, the meteoric theory of solar energy. In the aftermath of the Star's near collision with the earth, "the days were hotter than of yore" because the comet and the greater portion of the moon, having crashed into the sun, are conceived to have augmented not just its mass but also its energy, by the conversion of their mechanical energy upon collision into heat.

For these formulations of the first law in relation to solar energy, both Wells and Flammarion may have been indebted to the popular physical expositions of John Tyndall, professor of natural philosophy at the Royal

Institution from the 1850s to the 1880s and an avid popularizer of leading scientific research.[55] For instance, in the lead essay of his 1865 collection *Fragments of Science,* "The Constitution of Nature," Tyndall stated: "Acted on by [gravity], the earth, were it stopped in its orbit to-morrow, would rush towards, and finally combine with, the sun. Heat would also be developed by this collision . . . equal [to] that produced by the combustion of more than 5,000 worlds of solid coal, all this heat being generated at the instant of collision."[56] In the solar physics of Tyndall's time, decades prior to the understanding of nuclear energy, the sun's steady rather than diminishing production of light and heat was accounted for in several ways. According to the meteoric theory, the solar system supplied the sun a constant infusion of rapidly moving interplanetary matter, and this meteoric source of fuel augmented an allied theory of solar condensation, in which the sun maintained its energy at an even level through the constant shrinkage of its radius. During the debate at the Parisian Institute it is eventually pointed out that, catastrophic comets aside, in all probability the world will end when, its internal and external fuel supplies exhausted, the sun burns out:

> The probable cause of the heat of the sun is the condensation of the nebula in which this central body of our system had its origin. . . . The temperature of this great day-star does not seem to have fallen any; for its condensation is still going on, and it may make good the loss by radiation. Nevertheless . . . the time will come when the circulation, which now supplies the photosphere, and makes the central mass a reservoir of radiant energy, will be obstructed and will slacken. . . . The brilliant photosphere will be replaced by a dark opaque crust which will prevent all luminous radiation. . . . The earth, a dark ball, a frozen tomb, will continue to revolve about the black sun, traveling through an endless night and hurrying away with all the solar system into the abyss of space. *It is to the extinction of the sun that the earth will owe its death, twenty, perhaps forty million years hence.*[57]

But Flammarion holds this momentary reversion to entropic rationales in check by placing it in relation to a long list of more or less plausible doomsday scenarios, with the effect of diminishing their overall shock value. And in the immediate event, the comet deals the earth a severe but not irreparable global blow. So *La Fin du Monde* anticipates the simple structure of "The Star," an averted cosmic threat that produces a spiritual transformation of humanity. The latter part of Flammarion's lengthy didactic fable narrates our higher evolutionary destiny, the perfect mastery of material conditions. "Progress is the supreme law," the narrator declares, looking back upon a distant future when humanity "had risen . . . to the position of a sovereign who ruled the world by mind, and who had made it a paradise of happiness, of pure contemplation, of knowledge and of pleasure."[58] The future shape of human biological and social evolution produces a dated catalog of unremarkable marvels: the "weather could be predicted . . . electricity had taken the place of steam."[59] These technological developments are overshadowed by biological mutations that over many millennia unfold new powers of perception: "the electric sense . . . the psychic sense."[60] For millions of years humanity mounts evolution's rainbow to reach a hazy "apogee" of sentient development (fig. 13).

But now the world of the future is poised to fall back from its zenith. The time has come of inexorable physical decline brought on by the geological decrepitude of the earth. Many of the ecological disasters presented as physical hypotheses in the great comet debate of the twentieth-fifth century are now seen to come true. The first characters in the novel granted proper names are the last two survivors of the expiring posthuman race, Omegar and Eva, who are subjected to a humorously lugubrious apotheosis. One can sense the relief with which Flammarion disposed of this modicum of personal interest in order to proceed to the finale of his opus, the actual end of *La Fin du Monde* (fig. 14).

Flammarion brought his readers this far in order to assure them that science guarantees a new beginning after the end of the world. In the end, gravitation and the first law's guarantee of the mechanical conversion of motion into heat outmaneuver the second law of thermodynamics. The very headlong momentum of the comets that had threatened the earth partakes of the cosmic dynamism necessary to make a new start. For even "a planetary corpse, a dead and frozen world . . . still bears . . . within its bosom an unexpended energy—that of its motion of translation about the sun."[61] The meteoric theory of solar energy is now reprised in terms of the energies latent in the fall of planets into the sun: "If the earth should fall into the sun, it would make good for ninety-five years the actual loss of solar energy. . . . The fall of all the planets into the sun would produce heat enough to maintain the present rate of expenditure for about 46,000 years."[62] In fact, Flammarion is transcribing a table of nearly identical information offered in the final pages of Tyndall's magnum opus on thermodynamics, *Heat: A Mode of Motion*.[63] A few pages before this table, Tyndall had made a strong statement of the meteoric theory's positing of gravity over entropy:

> Whatever be the ultimate fate of this theory, it is a great thing to be able to state the conditions which certainly would produce a sun,—to be able to discern in the force of gravity, acting upon dark matter, the source from which the starry heavens *may* have been derived. For whether the sun be produced,

Fig. 13. L'Apogée. (The highest point.) From *La Fin du Monde.*

> and his emission maintained, by the collision of cosmical masses—whether . . . the internal heat of the earth be the residue of that developed by the impact of cold dark asteroids, or not—there cannot be a doubt as to the competence of the cause assigned to produce the effects ascribed to it. Solar light and solar heat lie latent in the force which pulls an apple to the ground.[64]

Tyndall's profession of cosmic optimism provoked a powerful counterargument from some thermodynamic doomsayers who envisioned the end of the world as an utterly final event. In a passage of *The Unseen Universe* we examined in the last chapter, Tait and Stewart replied directly to Tyndall's materialist declarations: "If the present physical laws remain long enough in operation, there will be (at immense intervals of time) mighty catastrophes due to the crashing together of defunct suns . . . growing in grandeur but diminishing in number till the exhaustion of energy is complete, and after that eternal rest, so far at least as visible motion is concerned."[65] This "eternal rest" awards the second law cosmological precedence over the first law, at least to the limits of the "seen" or material world, in order to credit the "unseen universe" of divine agency with the ultimate responsibility for bringing material worlds into being.

Tyndall ranked the first law over the second, in order to maintain a secular materialist position in which matter itself, aided by gravity, is endowed

Fig. 14. Épilogue: Après La Fin du Monde Terrestre—dissertation philosophique finale. (Epilogue: After the end of the world—final philosophical thesis.) From *La Fin du Monde.*

with the ability to regenerate energy. He stated this position most succinctly in his article "The Constitution of Nature." Extrapolating from the mechanical theory of heat to the dynamism of cosmic origins, Tyndall again affirmed: "In the attraction of gravity, therefore, acting upon non-luminous matter, we have a source of heat more powerful than could be derived from any terrestrial combustion. And were the matter of the universe thrown in cold detached fragments into space, and there abandoned to the mutual gravitation of its own parts, the collision of the fragments would in the end produce the fires of the stars."[66]

Flammarion extended Tyndall's promise of a self-organizing universe into an oscillating eternity of creations and destructions, envisioning in more elaborate detail Tyndall's intimation of the originary force of dark star crashes:

> Long after the death of the earth, of the giant planets and the central luminary, while our old and darkened sun was still speeding through boundless space, with its dead worlds on which terrestrial and planetary life had once engaged in the futile struggle for daily existence, another extinct sun, issuing from the depths of infinity, collided obliquely with it and brought it to rest!
>
> Then in the vast night of space, from the shock of these two mighty bodies was suddenly kindled a stupendous conflagration, and an immense gaseous nebula was formed. . . . [It] began to turn upon itself. And in the zones of condensation of this primordial star-mist, new worlds were born, as heretofore the earth was. . . . And these universes passed away in their turn.[67]

A consistent thermodynamic idiom of catastrophic and cataclysmic rhetoric runs from John Tyndall's popular expositions of the early 1860s to Flammarion's compendium of cosmological endgames and H. G. Wells's scientific fables, and from there to broad dissemination in the cosmological imaginary of international modernism. For instance, the "Ithaca" chapter of James Joyce's *Ulysses* (1922) tells of "the appearance of a star (1st magnitude) of exceeding brilliancy dominating by night and day (a new luminous sun generated by the collision and amalgamation in incandescence of two nonluminous exsuns)."[68] And again, as we noted in Yevgeny Zamyatin's essay "On Literature, Revolution, Entropy, and Other Matters" (1923), creative thermodynamic catastrophe was updated for application to current sociopolitical upheavals: "Two dead, dark stars collide with an inaudible, deafening crash and light a new star: this is revolution."[69] Joyce's nova and Zamyatin's vision of revolution have their sources in a specific recuperative Victorian moralization of the first law of thermodynamics in direct response to the heat-death scenarios elaborated on the basis of the second law. Spinning out of the cultural allegories constructed on these models of energy, the dark stars that rose over the discourse of classical thermodynamics heralded an explosive historical era driven in its economic and political transformations both by phantasmagoric specters of the entropic fates awaiting the world and its inhabitants, and by secular scientists' positive promises that new life can always be created out of violent collisions.

In these scientistic images of a cosmos variously doomed to imminent or ultimate energic ruin, the world is subjected to the accidental breakdown or inevitable obsolescence that typically befall the industrial commodity. In anticipation of contemporary responses to periodic energy crises, recuperative scenarios are then concocted by which the world is *recycled* and so salvaged out of its own junk heap. Flammarion counters his own entropic phantasmagorias by projecting an economic wish-image onto the physical universe. In the fullness of eternity, the commodified cosmos is granted the transcendental capacity to maintain itself in perpetually productive circulation. This ultimate guarantee retroactively underwrites the hopeful saga of posthuman evolution toward an apogee of happiness narrated earlier in the text. Walter Benjamin addressed his critique of modernity precisely to such scientistic myths of evolutionary progressivism. "The phantasmagoric understanding of modernity as a chain of events that leads with unbroken, historical continuity to the realization of social utopia, a 'heaven' of class harmony and material abundance" did not represent history's telos; rather "this conceptual constellation blocked revolutionary consciousness like an astrological force."[70] For Benjamin, the truly malign or baleful influence wrought by the dark stars of classical thermodynamics are the discursive collisions of these allegories of cosmic catastrophe with modern society. And yet, that dark astrology marked the beginning of the conceptual transition of thermodynamic ideas from the cycles of mythic history to the self-organizing possibilities of irreversible time.

Chapter 6
A Different Sun

Victorian physics calculated the age of the sun and determined the date of its demise in ways we now consider to have been very wide of the mark. In the first decades of the twentieth century, the modernist sun was poised between the prior regime of late-classical mechanism and determinism we examined in the last chapter, and the new physics of atomic energy and quantum indeterminacy. During this period of conceptual limbo, nuclear gradually replaced chemical explanations for the operation of stars. The sun we see today is neither a scientific nor mythic absolute but a phenomenon relative to the shifting explanatory frames of its observers.

The writings of D. H. Lawrence and Yevgeny Zamyatin capture this transitional phase of perceptual relativity through revisionary allegories of *a different sun*. At the same time, residual conceptual investments in classical thermodynamic paradigms drove both writers to continue formulating antimechanical rather than postmechanical schemes for the sun's reprieve from its former entropic death sentence. In curiously similar ways, both Lawrence and Zamyatin articulated mythopoetic terms for this cosmological release by refracting the sun into multiple reference frames and then reading this refraction through vitalist rhetorics of unforeseeable or insurgent renewals. Their visionary scientistic defenses against the abiding cultural authority of mechanistic positivism mark the continuity as well as the distance between the secular or mystic materialism of Tyndall and Flammarion and the new complementarities of relativistic modernism.

Zamyatin and Lawrence have seldom been examined before in each other's proximity. Between them they epitomize the concurrent reactions of European and Anglo-American culture to modernist crises in scientific authority. While both resisted late-classical scientisms, in both classical thermodynamic rationales remained powerfully interwoven with biological discourses of evolution, degeneration, and vitalism.

Zamyatin's novel *We* has long been acclaimed as a science-fiction classic, and his critical writings explicitly invoke thermodynamic principles in relation to cultural arguments about aesthetic and social renovation. Thus his writings lend themselves directly to a discussion of allegory and science. The case of Lawrence is somewhat more oblique. He has often been stereotyped as an avatar of Romantic irrationalism and antiscientific sentiment. Nor has he been broadly appreciated as an allegorist per se, but rather as a late-romantic symbolist. Lawrence's work was importantly driven by reactions against "science," but for this reason it is especially important to be clear about *which* science, with *what* specific cultural forms and effects, he was reacting against. Historical precision about these matters opens up Lawrence for his allegorical as well as symbolist dimensions. In his and Zamyatin's discursive and novelistic adaptations of Plato's cave imagery, for instance, it is precisely the mechanistic determinism of late-classical science that keeps us imprisoned in rather than springing us out of a state of existential ignorance. Comparable perceptions of the role of science in the modern struggle between doctrinal dogmatism and spiritual liberation energize the high mythopoetic manner of both writers.

Order and Chaos in *We*

Zamyatin's *We* was composed at the turn of the 1920s, in the extraordinary moment of early postrevolutionary Russia. It is both a premier example of literary allegory in critique of shifting doctrinal regimes and a classic satire of technoscientific modernity. We will first examine the zero degree of its literary structure—the way *We* establishes its allegorical temper on its own scriptive processes—and then describe the energic components of *We*'s historical density.

Scenes of Writing

The text presents itself as OneState Number D-503's journal, written a thousand years in the future, begun in patriotic response to a call for papers to be launched into space on the maiden voyage of the INTEGRAL, One-State's first outer-space vehicle. "Everyone who feels himself capable of doing so," D-503 reads in the *State Gazette,* "is required to compose treatises, epic poems, manifestos, odes, or other compositions dealing with the beauty and grandeur of OneState."[1] Carrying a payload of public testimonials, this interplanetary herald will be an angel bearing glad tidings of OneState's perfection to the planets and stars. *We* is set into motion against the INTEGRAL's cosmic tableau of daemonic communication—technogods carrying imperial messages from star to star—and D-503 begins to write in sublime identification with this angelic prospect. This device situates readers of

We among an unknown and literally alien audience in need of detailed interpretive commentary. The builder of the INTEGRAL, D-503 is exhilarated at the prospect of stocking his space vehicle with scripts, seeding its nucleus with a written treatise on the means and significance of its own construction.

The very medium of allegory is specular, self-reflexively turned toward its own verbal construction. As a narrative embedded in this mode, *We* orbits its own scenes of writing.[2] D-503's records are both digressive, a nonlinear series of thematic riffs, and increasingly tentative, as an unknown and unaccountable chaos invades his previously deterministic existence. The proud document he had envisioned turns into a madman's diary. Only his commitment to unflinching scientific veracity before his interplanetary jury keeps D-503's literary project going. We participate in D-503's disorientation and loss of absolute reference points, becoming unsure ourselves whether the text in hand is virtually consecutive in OneState time or a random sheaf of world-lines. The vigorously modernist, unreliably jumbled text of the novel reflects the daemonic moral of D-503's uncanny journey out of OneState certainties, ultimately uttered by the Mephi conspirator I-330: "Who knows who you really are? A person is like a novel: Up to the very last page you don't know how it's going to end. Otherwise there'd be no point in reading."[3]

Zamyatin constructs allegorical meanings across the gaps in the temporal and spatial dimensions of D-503's worldview. The text results from explanations to the distant and unknown audience the narrator hopes to encounter on some alien world in the future. At the same time, this well-indoctrinated engineer pictures his own past through state-sanctioned images of his primitive progenitors—Zamyatin's virtual readers: the benighted denizens of the twentieth century. The temporal gap between OneState man and his bestial ancestors reemerges within the psychological space that separates D-503's proper social identity as a prime cog in the OneState machine from his obsolete, individual organic endowments: "I can't stand people looking at my hands. They're hairy, shaggy, some kind of stupid throwback."[4] D-503's initial arrogance toward "the old God and the old life" (7) is another symptom of the positivist and evolutionist scientisms under Zamyatin's critique. "Science is going forward" (9), D-503 exclaims, and the icon of this truth is the creation of the INTEGRAL. Zamyatin's satire of early-twentieth-century military-industrial technoscientism pivots on this space ship. Like a pillar of sacrificial fire, it towers in the background of the novel, dedicated to the totalizing colonial ambitions of OneState. D-503 comes forward inscribed with the tensions produced when positivist technoscientism and political totalitarianism meet under the banner of social evolutionism.

But as he prepares to launch triumphant messages out to other worlds, his own suffers a total metamorphosis. *We* is constructed through D-503's self-analytical written confessions of the breakdown of his cosmos, the crash that occurs when the well-ordered universe of OneState science collides with the mythic forces marshaled by the Mephi underground.

Titans of Scientism

In the first decades of the twentieth century, classical thermodynamics remained a significant source of technoscientistic verities and anxieties.[5] However, by the 1920s Zamyatin could observe that even while classical thermodynamic scientisms retained broad social authority, the deterministic underpinnings of the classical science itself were becoming theoretically inadequate. In *We* Zamyatin satirizes these ironies of conservative scientism, the ways that cultural investments are maintained in the progressive authority of particular scientific enterprises long after their theoretical foundations are discarded, superseded, or otherwise displaced from epistemological preeminence. Three years after completing *We*, Zamyatin codified his own literary scientism in "On Literature, Revolution, Entropy, and Other Matters." Here he championed recent advances in physics and other current challenges to the mechanistic materialism of classical thermodynamics: the non-Euclidean geometries of relativity theory and the nonmechanical behaviors of wave and particle physics. But his manifesto was equally classical in thermodynamic orientation. A clash of "new" and "old" scientific testaments, replicating the doctrinal confrontations of religious allegories, animates both Zamyatin's narrative fiction and his critical sequels.

The cosmos of OneState is layered with an overlapping collage of the technoscientistic discourses mediating industrial modernity's various ideological needs. For instance, the state religion of OneState is modeled on the syncretic scientism of late-classical mechanism. *We* submits to critical satire the fulfillment of the most radical projects of mechanistic and positivistic technopolitics at the start of the twentieth century. The thirtieth-century social engineering of OneState fictively realizes the totalized efficiencies of late-classical positivist scientisms driven by nineteenth-century advances in physics—the mechanistic formations of classical thermodynamics—as well as the concurrent schemes of evolutionist technocrats. OneState's monumental bureaucracy extrapolates from positivist thermodynamicism to a centralized social energetics, and from social evolutionism to a totalized eugenics. Merged into OneState's marching masses, D-503 wraps an atavistic protofascistic imagery of "Assyrian" formations around his vista of this technoscientistic millennium. *We* follows *The Time Machine* in subverting rosy anticipations of either proletarian or bourgeois millennia. The progressive perfection of the total social machine is instead a savage lapse into monumental theocratic statehood. Zamyatin's narrative unfolds a prime blueprint for dystopian technopolitical allegory in the twentieth century.[6]

Zamyatin reprises an important version of physical and mathematical scientism with the OneState dogma of the "happiness fraction": "Bliss and envy are the numerator and denominator of that fraction known as happiness. . . . The denominator of the happiness fraction has been reduced to zero and the fraction becomes magnificent infinity."[7] The actual inventor of a thermodynamic happiness formula was Wilhelm Ostwald, the German pioneer of physical chemistry who from 1890 onwards developed a "generalized thermodynamics" under the name *energetics.*[8] In Ostwald's scheme of *Energetik,* with the dynamics of matter, life, and mind all resolved into energy transformations, the first and second laws of thermodynamics circulated freely between the physical world and the ethical sphere. Zamyatin's "happiness fraction" is a satirical variation of Ostwald's ethical scientism.

"The heart of Ostwald's ethics is the 'energetical imperative' which says 'Vergeude keine Energie, verwerte sie' (do not waste energy: make it useful). . . . For him the basic ethical principle derives directly from the second law of thermodynamics" in its energetic extension.[9] To judge from his happiness formula in particular, for Ostwald the moral imperative of social energetics is to reduce the frictions generated by resistances to individual wills. Happiness is self-determination without external check. As Casper Hakfoort notes, the "remarkable thing about Ostwald's formula is that happiness is determined entirely by quantities of energy":

> $$G = (E + W)\,(E - W),$$
>
> where G [*Glück*] represents the quantity of happiness. E [*Energie*] is the energy applied in accordance with the individual's will, and W [*Widerstand,* resistance] is the energy exerted unwillingly. The factor $(E + W)$ gives the total energy, and Ostwald's supposition is that this quantity is large in youth and small in later life. In his view, the factor $(E - W)$ allows for the fact that, where there is opposition or resistance from others (costing an amount of energy W to overcome), happiness is diminished. (535)

This moral algebra is arithmetically straightforward. If, to accomplish a given goal, one must exert equal amounts of willing energy E and unwilling energy *W*, one experiences a null result, neither pleasure nor unpleasure. Assigning a unit value to each, the formula reads: (1 + 1) (1 – 1) = (2) (0) = 0. Whenever one exerts greater energy of resistance *W* than energy of desire *E*, the second term and thus the full product of the formula is always negative; for instance, expending twice as much *W* as *E* results in three units of unpleasure: (1 + 2) (1 – 2) = (3) (–1) = –3. Things are symmetrically reversed when *E* is greater than *W*: twice the desire over the resistance yields three units of pleasure: (2 + 1) (2 – 1) = (3) (1) = 3. Ostwald's quantity of unwilling energy *W* is a scientistic translation of entropy into a voluntaristic ethics: *W* measures wasted or dissipated energy of desire, energy lost to the uses of the will. By the same token, in the quantity *E*, energy per se becomes (re)identified with human spiritual purposes, and the old ethical polarities of free will and determinism are replayed as a ratio of energy and entropy.

However, by another simple reversal, thermodynamic scientism can also stigmatize free will as entropic waste and valorize regimented behavior as energetic effectiveness. This is the case with the "scientific management" developed by American efficiency expert Frederick Winslow Taylor. *We*'s parodic elevation of Taylor to a position of supreme philosophical and cultural authority in OneState recalls a related movement of technoscientistic hegemony at the beginning of the twentieth century. Like Ostwald's *Energetik,* Taylorism developed into a form of scientized moralism, but here the emphasis was distinctly corporatistic rather than individualistic. Taylor's ethical reflections on energy forms propelled a cluster of cultural organizations and modernization crusades.[10] And Zamyatin's transformation of Taylor into a patron saint of OneState was no idiosyncratic choice. Richard Stites has documented a sizable Soviet vogue for Taylor in the second and third decades of the century, when numerous movements, "efficiency crazes," sought Taylorist solutions to industrial and social reorganization.[11] Taylorism was a significant part of the larger movement of economic and cultural rationalization across the industrialized and industrializing nations, embraced by capitalists and commissars as a hallmark of modernization, a progressive technique for optimizing economic productivity.

Taylorism in *We* is an important emblem of thermodynamic allegory, a powerful metonymy for the scientistic breadth of productivist anxieties over the exhaustion of energy resources. For a capitalist accounting, the economic toll of mechanical dissipation was registered most immediately on the bottom line of a corporate balance sheet. To maximize profits as well as forestall the heat death of a heavy-industrial civilization hell-bent on world domination, Taylor combated entropy at the spatiotemporal interface of flesh and metal, the contact point between the physiological energies of human labor and the engine-driven implements of mechanized industries.[12] Taylor's systems of time-and-motion management attacked wasted energy through a pervasive extension of thermodynamic principles to general physiology and human psychology.

Taylorism's top-down centralization of industrial processes provides Zamyatin with an ideal model for the political state as an optimized mechanism for the enforced rationalization of all human movements. Social and mental order and disorder are moralized as the optimized working and the breakdown of a monumental machine. From the vantage of the "scientific manager," the Taylorist elimination of individual differences throughout a deliberate sequence of collective bodily movements produces a mechanical choreography whose aesthetic power overrides the dehumanization of the dance, even while its efficiencies yield a gain in productivity. D-503's technological aesthetic parodies the political excesses of Futurism as well as the "Taylorist art movements" that would spring up throughout the USSR in the 1920s. The Taylorite dance of myriad identical units simultaneously performing identically repeated motions gives Zamyatin a panoramic image for a productivist aesthetics of totalitarian statism. D-503 is an accomplished connoisseur of this cult of industrial form, whose transports of aesthetic feeling reveal the phantasmagoric mechanisms of his ideological indoctrination:

> Just this morning I was at the hangar where the INTEGRAL is being built—and suddenly I caught sight of the equipment: the regulator globes, their eyes closed, oblivious, were twirling round; the cranks were glistening and bending to the left and right; the balance beam was proudly heaving its shoulders; the bit of the router was squatting athletically to the beat of some unheard music. I suddenly saw the whole beauty of this grandiose mechanical ballet, flooded with the light of the lovely blue-eyed sun.[13]

D-503's aestheticized vision of OneState's productivist determinism turns from the beautiful to the sublime when he contemplates the Taylorist bible represented by the Table of Hours, which "turns each one of us right there in broad daylight into a steel six-wheeled epic hero" (13). Sublime identification powers the political mechanisms through which OneState Numbers participate in the dynamical infinity of the whole population. These aesthetics of regimented beauty and totalitarian sublimity inhabit a political organization resting directly on the mechanical determinism of classical thermodynamics.

However, the destination of Taylorist aesthetics is distinctly anaesthetic. The ultimate scientistic solution, the final positivist fix at the end of this fiction, is the numb bliss of "fantasectomy."[14] OneState reacts to the revolt of the Mephi by instituting a mass Operation that irradiates the brain in order to cauterize the soul. Zapping the last vestiges of free will's wasted energies from the human organism, the Operation will rub the frictions of human unpredictability out of the social mechanism. The *State Gazette*'s announcement of the Operation reads: "The latest discovery of State Science: The imagination is centered in a wretched little brain node in the region of the *pons Varolii.* Expose this node to three doses of X rays—and you are cured of imagination."[15] Previously OneState collectivity had bowed down to the hieratic execution Machine that, rising high above the human populace, vaporized criminal Numbers. The Operation now obsolesces the Machine by consummating the total mechanization of humanity. With this dark satire *We* presents a striking summation of the lethal mechanisms threatening industrial societies from the total social abuse of classical thermodynamic scientisms.

We remarked in chapter 2 that classical thermodynamic discourses produce a commodification of the cosmos by picturing the universe as an industrial machine driven by solar heat engines. Taylorism and other sciences of work imputed to the microcosm a parallel commodification, that is, they inscribed on the organic body as well as the cosmos the instrumental economies of industrial machines. Bringing sexual reproduction under the mechanical efficiencies of commodity production moves thermodynamic scientism into the range of social evolutionism.[16] *We* dramatizes the ideological violence inhabiting social evolutionist discourses that glorify the present by disfiguring the historical past. The dogmatic allegories of OneState scientism have schooled D-503 to enjoy a superb contempt of the Ancients. He denigrates the cloudiness of their poets, the chaos of their cities, the murderous "condition known as freedom" (15). In newly postrevolutionary societies like the Soviet Union of 1920, where the ruins of despised old forms still clutter the present landscape, the demonization of defeated and forsaken forms of culture has a special rhetorical appeal. But that discursive tactic exists wherever cultural or political authority legitimates itself by demonizing some phase of the past as an era of pitiful ineptitude or active abomination. Such demonization turns history's victims into allegorical caricatures, icons of evil.

We plots the cultural tensions that contributed to the wider modernist impasse between thermodynamic degradation and evolutionary individuation. This interaction of phantasmagoric energy forms—entropy and the life force—produced an imagery of oscillation, itself scientized as the unstable or incessant reversal of electromagnetic polarities. D-503's narrative presents a series of dynamical polarities as both scientistic symptoms and psychological motives. He experiences sudden lurches from positive to negative extremes, a clash of opposites without dialectical sublation or synthesis and without peaceful political solution. One of his early Records documents that after a moment of sublime identification with the evolutionary perfection of OneState, "there came a moment, a leap across the centuries, from + to –.I recalled (association by contrast, apparently), I suddenly recalled a picture in the museum: one of the avenues they had back then, after twenty centuries—a stunningly garish, mixed-up crush."[17] Here the oscillating poles are graphed on a temporal axis broken into the conquering present and a not-entirely-vanquished past. As D-503 falls from myth into history, from symbolic consubstantiality into allegorical temporality, the disjunctions of evolutionary time challenge the notion of OneState's achieved perfection. At the same time, polar reversals structure D-503's own oscillating lapses of identity, as his ideological identifications break down under the subversive pressures of the Mephi uprising. These political polarities destabilize his ability to attach a univocal meaning to the *we* to whom he belongs: "'They' . . . 'We' . . . How am I to know?" (141).

The polarisms of the text significantly resolve into plus and minus positions on a temperature scale, the thermodynamic polarity of hot and cold. OneState's happiness fraction, says I-330, is the cultural equivalent of heat death: "'What a stupid prejudice it's been all these years to put a plus sign in front of happiness. Absolute happiness should of course have a minus sign, a divine minus.' I remember muttering distractedly: 'Absolute minus is 273°.'"[18] In this milieu the sun becomes an oscillating icon drawn into the circuits of evolutionary and emotional as well as thermodynamic polarity. In *We* the sign of the sun is a complex node of allegorical implications. Zamyatin divides the sun between D-503's crystal blue thermodynamic orb and I-330's throbbing red vitalist flame. D-503 first experiences I-330's seductive snakebite as a solar hallucination: "Not this one, not ours, shining all sky-blue crystal regularity through the glass brick—no: a savage, rushing, burning sun—flinging everything away from itself—everything in little pieces" (18). In the charged figures of this metamorphic sun Zamyatin compresses the phantasmagoric play of energy forms that drives the novel. The sun gathers together the author's allegorical clash of scientistic titans.

In its more typical thermodynamic register, Zamyatin figures the cooling or entropic sun in shades of blue. D-503 accepts this icy or crystalline sun as the norm of solar perfection, shining a hard pale emblematical light down upon the efficient social mechanism of OneState. The specifically thermodynamic signification of the blue sun is confirmed later when D-503 experiences his sexual failure with O-90 as a kind of cosmic chill. Zamyatin counters the pallid and cooling entropic sun of classical thermodynamics with its designated allegorical counterpart, the youthful, blood red or living sun of evolutionary vitalism. In erotic delirium over I-330, D-503 enjoys a vision of living energies in which the morning sun is "rosy, transparent, warm gold. And even the air is a little pink, and everything's saturated with the tender blood of the sun, everything's alive; soft and alive—the stones; warm and alive—the iron; the people alive and everyone of them smiling."[19]

The sun as the emblem of a living or vitalist cosmos unbroken by the threat of entropic finalities is a part of Zamyatin's wider treatment of biological scientisms. He divides this treatment between the two main female characters. O-90 centers the novel's treatment of eugenic schemes, while I-330 represents a form of erotic vitalism linked to the larger sexual revolutionism of the Mephi. These two allegories converge as O-90 liberates her maternal desire and falls in with the Mephi herself in rebellion against sexual regimentation. D-503 considers OneState's synthesis of eugenic sexual regulation and Taylorist social engineering to be the primary motor of OneState's great felicity, its grand solution for the happiness formula. The *Lex sexualis* of OneState follows from the premise of the absolute interchangeability of the Numbers.

The progress of D-503's and O-90's liberation from the "beneficent yoke" of OneState leads precisely to a revival of egoistic individuality with a vitalist reproductive agenda, acutely registered by the pains of envy and jealousy that energize their love triangle. This is a consummate synthesis of the anarchistic sexual progressivisms that inhabited early modernism, confronting an equally typical cluster of authoritarian positivist scientisms. This collision is another example of the great modernist trainwreck, the cosmic crash unfolded out of classical thermodynamics from the seeming insolubility of the polarities separating physics and biology, energy and life, entropy and evolution, order and randomness, the machine and the organism. As D-503 phrases it at the climax of his revolutionary crucifixion: "My head was going in all directions. Two logical trains had crashed, piling on top of each other, crumpling, flying apart."[20]

Eden as Equilibrium

We subverts a utopian tradition founded on the myth of Eden.[21] Traditional utopias are purified states of perpetual isolation, paraspaces hermetically sealed from intrusion. Under a thermodynamic regime, however, closed spaces are precipitated into time and subjected to the rule of increasing entropy. All Edens become entropic, spaces of inertia in need of revolutionary reenergizing. *We* shows that classical thermodynamics has been instrumental in redefining utopian enclosure as dystopian confinement.

Classical thermodynamics is strictly conceptualized on the basis of the environmental closure of the system. According to the second law, isolated physical systems evolve toward thermal equilibrium. In the 1870s Ludwig Boltzmann pioneered the use of probability theory in order to describe the mechanism of entropy-increase toward equilibrium as the drift toward a maximally probable state. Boltzmann's statistical mechanics were based on previous work by Maxwell, who brought into his physical calculations mathematical population models developed in the 1830s by the French sociologist Quetelet.[22] This cultural flow of ideas from social into physical science inscribed thermodynamics at its source, as a science of molecular populations and their "tendencies," with a sociological subtext enabling physical systems to be moralized according to the statistical behaviors of human masses. These lines of conceptual connection in the nineteenth century laid an allegorical groundwork, a subtextual readiness, for the sociocultural moralizations of thermodynamics that then spill over into the twentieth century.

Ilya Prigogine and Isabelle Stengers provide a helpful synopsis of Boltzmann's innovations, as well as a discussion of equilibrium thermodynamics that sheds light directly on the satire of classical thermodynamics abroad in *We*. They expound Boltzmann's use of probability in forming his "order principle" by stipulating an apparatus highly reminiscent of Maxwell's Demon scenario: "an ensemble composed of *N* particles is contained in a box divided into two equal compartments."[23] But instead of the diaphragm with weightless slide door by which Maxwell divided *A* and *B*, here compartments 1 and 2 are not separated from each other. They simply mark the equal division of a unitary space. Given such a closed ensemble of *N* elements, the total number of possible arrangements of those elements will be *P*, most of them probable, some of them unlikely. "Dissymmetry" refers here to an initially *improbable* distribution, a statistically unlikely ordering of elements, a *low-entropy* state.

With regard to the laws of probability, what we want to know is this: of the set of *N* particles, how many "complexions" (*P*)—or different ways of distributing the particles between the two compartments—exist, and what are the relative probabilities of finding the total system in any particular distribution? Combinatorial analysis provides a basic formula for answering such questions:

$$P = N! \div (N_1!)\,(N_2!),$$

where $N!$ is a factorial number; N_1 represents the quantity of particles in compartment 1; and N_2 is the remainder of particles in compartment 2. Factorial numbers work this way:

1!	$1 \times 1 = 1$
2!	$1 \times 2 = 2$
3!	$1 \times 2 \times 3 = 6$
4!	$1 \times 2 \times 3 \times 4 = 24$
. . .	
8!	$1 \times 2 \times 3 \times 4 \times 5 \times 6 \times 7 \times 8 = 40320.$

For example, if there are 8 particles distributed in the total system composed of compartments 1 and 2 ($N = 8$), then what is the number of complexions in which all 8 particles occupy only one of the two compartments? That is, how likely is the most highly ordered case?

$$P = 8! \div 8! = 1.$$

In contrast to this extreme case, how many complexions yield a distribution in which the particles are equally divided between both compartments?

$$P = 8! \div (4!)\,(4!) = 70.$$

It is thus, in this system, seventy times more likely to encounter an equal distribution than the most unequal distribution. For most real-world systems the value of N is vastly higher. Prigogine and Stengers give 10^{23} as an average, and "for systems composed of a large number of particles, all states that differ from the state corresponding to an equal distribution are thus highly improbable."[24]

> Boltzmann was the first to realize that irreversible increase in entropy could be considered as the expression of a growing molecular disorder, of the gradual forgetting of any initial dissymmetry, since dissymmetry decreases the number of complexions when compared to the state corresponding to the maximum of P. Boltzmann thus aimed to identify entropy S with the number of complexions: entropy characterizes each macroscopic state in terms of the number of ways of achieving this state. Boltzmann's famous equation $S = k \lg P$ expresses this idea in quantitative form. . . .

> Boltzmann's order principle implies that the most probable state available to a system is the one in which the multitude of events taking place simultaneously in the system *compensates for one another statistically*. In the case of our first example, whatever the initial distribution, the system's evolution will ultimately lead it to the equal distribution $N_1 = N_2$. (123–24)

The distribution $N_1 = N_2$ indicates that the system has evolved from an unequal or dissymmetrical complexion to a state of statistical equilibrium. In other words, its entropy has increased to a maximum.

However, there are many routes to equilibrium states, some of which result in a *lowering* of entropy. That is, a state of "statistical compensation" in which the competing energies of a system lock into the balance of equilibrium can also occur in highly ordered structures, such as crystals. In this case, the equilibrium of the structure expresses a minimum of free energy rather than a maximum of entropy: "In a closed system, defined by boundary conditions such that its temperature T is kept constant by heat exchange with the environment, equilibrium is not defined in terms of maximum entropy but in terms of the minimum of a similar function, free energy: $F = E - TS$, where E is the energy of the system and T is the temperature (measured on the so-called Kelvin scale, where the freezing point of water is 273°K and its boiling point is 373°K). This formula signifies that equilibrium is the result of competition between energy and entropy" (125–26). Under the regime of classical thermodynamics, which posits the equilibrium of a closed system as the norm, *energy and entropy are already and inevitably conceptualized as antagonists*, "forces" whose struggle can be delayed or diminished, but out which can come only a minimum of free energy or a maximum of disorder.

Prigogine and Stengers's further discussion of equilibrium structures in physics provides a precise rehearsal of the itinerary of Zamyatin's sociopolitical allegory in *We*. In that allegory, the thermodynamics of molecules in closed systems at equilibrium is the physical analogue of the political status of individuals in a totalitarian regime, within which one is either imprisoned by order or liberated by chaos. "In order to produce equilibrium, a system must be 'protected' from the fluxes that compose nature,"[25] and so OneState has crystalized the Green Wall intended to seal itself in from its natural surroundings.[26] The cool blue tones and translucent ideals of OneState respond to the fact that for a closed physical system, the most effective way to lower entropy and maintain order is literally to lower the temperature. In the competition between energy and entropy, "temperature is what determines the relative weight of the two factors":

> At low temperatures, energy prevails, and we have the formation of *ordered* (weak-entropy) and *low-energy* structures such as crystals. Inside these structures each molecule interacts with its neighbors, and the kinetic energy involved is small compared with the potential energy that results from the interactions of each molecule with its neighbors. We can imagine each particle as imprisoned by its interactions with its neighbors. At high temperatures, however, entropy is dominant and so is molecular disorder. The importance of relative motion increases, and the regularity of the crystal is disrupted.[27]

Crystals maintain molecular order by "imprisoning" the kinetic energies of their elements in molecular bonds of great stability. The "free energies" of unbonded molecules are sacrificed to the collective order of the molecular structure, and this energic equilibrium is maintained by a low-temperature regime offering minimal perturbations to the system. However, the infusion of energy represented by high temperatures eventually leads only to the other form of equilibrium, the averaging out of heat energies in the state of thermal equilibrium. The great irony and equivocation of equilibrium structures is that the statistical averaging that leads to equilibrium operates at both extremes of order and disorder. The unlikely but ordered equilibrium of the low-entropy crystal, despite its aesthetic appeal, is just as inert as the disordered equilibrium of a highly entropic system in the dull condition of maximum thermal probability. "By definition," Prigogine and Stengers continue, equilibrium structures "are inert at the global level. For this reason they are also 'immortal.' Once they have been formed, they may be isolated and maintained indefinitely without further interaction with their environment. When we examine a biological cell or a city, however, the situation is quite different: not only are these systems open, but also they exist only because they are open."[28]

Thus any direct extrapolation from classical thermodynamic descriptions will predetermine the kind of evolutionary developments described by biology and the social sciences as rare and unpredictable aberrations. Throughout this period the only evolutionary attractor that could be deduced from the fundamental laws of physics was the approach to thermal equilibrium, the inevitable and ultimate energic state of any closed system: "In many respects, it is the least interesting state of matter: for any living organism it corresponds to death. But for well over a century—since its origins in the middle of the nineteenth century—most physical scientists have concentrated their efforts on understanding matter in this dreary state."[29]

Within the equilibrium regime of an environmentally closed system, like a pendulum running down, what dynamism the system still possesses is played out in an oscillating dynamic of polarized energy forms. The interplay of kinetic and potential energies is itself opposed by the "forces" of entropy that attract the system toward an end point of enervated inertia. In the classical thermodynamical context, then, the privileging of equilibrium states creates a structural tendency to conceptualize the relation of energy and entropy according to a simple inverse ratio, a zero-sum process whose oscillations die down into inert equilibrium. In contrast, contemporary science constructs a more complex picture of an asymmetrical relation of collaboration as well as a conflict between energy and entropy. Equilibrium is death just because *imbalance*—the far-from-equilibrium status of "dissipative" structures—is necessary to make entropy productive.

We still inhabits the classical thermodynamic era: it posits an equilibrium model and the unstable but inexorable wobble of polar opposites as a universal condition. Consequently, insofar as we take I-330 to express Zamyatin's own theoretical orientation, his text succumbs to a noninevitable polarizing of energy and entropy that leads to images of impasse and conflict seemingly resolvable only by trainwrecks and explosions.[30] Nowhere is this clearer than in the scientized political lectures I-330 delivers to D-503. On one occasion, she expresses universal thermodynamic reality precisely as a dualistic polarization: "Look—there are two forces in the world, entropy and energy. One of them leads to blissful tranquillity, to happy equilibrium. The other leads to the disruption of equilibrium, to the torment of perpetual movement."[31] Shortly thereafter she debates D-503 again on the meaning of social and political perfection. Taunting her lover, she demands to know what comes after the total perfection of the state: "Then what?" Still an unreconstructed advocate of happy equilibrium, D-503 retorts: "Then nothing! Finished. All through the universe, evenly spread, everywhere":

> "Aha—evenly, everywhere. That is precisely what we're talking about—entropy, psychological entropy. You're a mathematician. Surely you see that only differences, differences of temperature, only contrasts in degree of heat, only that makes for life? And if throughout the universe all bodies are equally warm, or equally cool . . . You've got to smash them into each other—so there'll be fire, explosion, inferno. And we—we're going to smash them."[32]

I-330's infernal wisdom is a late reprise of the dark star crashes that Tyndall and Flammarion had earlier envisioned as the mode of cosmic collision through which the first law recuperated the second law of thermodynamics. Zamyatin marked his own subscription to this classical scenario several years later in "On Literature, Revolution, Entropy, and Other Matters," when he pictured the revolutionary renovation of society according to the same model: "Two dead, dark stars collide with an inaudible, deafening crash and light a new star: this is revolution."[33] Yet despite the glamor of this idiom of thermodynamic defiance, it remains socially symptomatic not of constructive resolution or even productive imbalance but of interminable impasse.

One is tempted to say that the modern world needed to get past the impasse of classical thermodynamics before it could begin to get beyond the divisiveness of purist ideological separatisms. Be that as it may, in the line from an old blues number, Zamyatin had "the right key, but the wrong keyhole." Through a brilliant allegorical deployment of classical thermodynamic analogues, his narrative elaborated an impressive systems model of society. But he was saddled with the wrong kind of system. One would now favor, from among physical models, Ilya Prigogine's description of dissipative or far-from-equilibrium structures, and from among sociological models, the autopoietic model of operationally closed but environmentally open social systems.[34] However, Zamyatin's world lacked these choices, and the environmental closure imposed on his text by the equilibrium model of classical thermodynamic systems foreclosed OneState's social options.

Lawrence's Cosmological Fantasia

> Men may be utterly different from the things they now seem. And then they will behold, to their astonishment, that the sun is absolutely different from the thing they now see, and that they call "sun."

The sun is such a prominent figure in the culture of classical thermodynamics because of its crucial yet equivocal double inscription in the registers of energy *and* life, physics *and* biology, at a time when the scientistic effects of these rapidly modernizing disciplines were already strenuously intertwined. In parallel with Yevgeny Zamyatin's narrative and discursive fictions, D. H. Lawrence's mythopoetic responses to the mechanized cosmos of Victorian thermodynamics exhibit a similar clash of modern scientisms. Lawrence's revisionary cosmology counters institutionalized cultural narratives of energy and evolution with vitalistic adaptations of electromagnetism.[35] His attempts to get beyond Victorian thermodynamics through an electrovital refurbishing of Romantic physiology also had mixed results, leading both to doctrinal impasses and liberating passages.

The Crown

Conceiving of life as a force or energy, Lawrence developed his diagnoses of modern morbidity as the social suppression of that life-energy in an imagery of subterranean spaces of hermetic enclosure and spiritual subjection. His major doctrinal statement *The Crown,* drafted in 1915 and amended in 1925, reworked Plato's parable of the cave in describing contemporary humanity not as chained down but still "enclosed within the womb of our era . . . the walls of which are shut, and have no window to inform us that we are in prison. We roam within the vast walls of the womb."[36] The existential impasse of the modern era here is a horrific constriction from which, for most persons, no vital delivery seems to be possible. In some important 1925 revisions to *The Crown,* Lawrence further applied Platonic allegorical motifs to condemn a series of Victorian and Edwardian scientific dogmas as stifling existential enclosures. His critique concerned the way that these fallible speculative systems presented their axioms as accomplished finalities, absolute truths, in short, as irrevocable laws:

> So, if we are imprisoned within walls of accomplished fact, experience, or knowledge, we are imprisoned indeed. The living sun is shut out finally. A false sun, like a lamp, shines.
>
> All absolutes are prison-walls. These "laws" which science has invented, like conservation of energy, indestructibility of matter, gravitation, the will-to-live, survival of the fittest: and even these absolute facts, like—the earth goes around the sun—or the doubtful atoms, electrons, or ether—they are all prison-walls, unless we realise that we don't know what they mean.[37]

Lawrence would deflate the pride of the modern intellect by opening it to further, intuitional meanings, by reinstating a sense of awe before the ultimate mystery or unfathomability of a sentient body with a "biological psyche."[38] But also, and more concretely, Lawrence was aware that many scientific concepts—like the luminiferous ether of late-classical wave theory—were tentative theoretical models of conceivable functioning for phenomena that are strictly speaking impossible to know through any immediate sensory experience. In his own way and to a certain degree, Lawrence participated in the revolution in scientific thinking that has displaced the late-classical, mechanistic physics in which he and his generation were schooled, and which was being dismantled and superseded throughout his active career.[39] At the same time, again like Zamyatin, much of Lawrence's sensibility remained tied to forms of Romantic and Victorian scientism. In particular confrontation with the doomsday moralizations of classical thermodynamics, Lawrence improvised a corpus of cosmological myths specifically to attack ideological structures imposed upon his culture by the unregulated dissemination or unwarranted acceptance of prior scientific fables.

In the passage from *The Crown* cited above, the provenance of thermodynamic ideas is announced first of all by Lawrence's figuration of the sun. As authorized physics was establishing the nuclear rather than chemical nature of the sun's energy, Lawrence constructed his own scientific romance or scientistic allegory of abiding solar energy from the vitalistic premise of a "living sun" that is only "shut out" by false knowledge. The sun is a symbol primevally marked with mythic inscriptions, but historically considered, it becomes such a pivotal figure in Lawrence's modernist myth-making precisely because Victorian physics had diminished its existential substance by envisioning it as a waning heat engine. Animating the sun and inscribing it within his vision of a vitalistic cosmos, Lawrence's mythopoesis here is another visionary scientism developed in response to the cultural dissemination of the entropic scenarios of the later nineteenth century. Lawrence's determinations of a "living sun" construct a counterallegory in defiance of anachronistic yet still dominant social myths of heat death.

As we have seen, for a number of leading Victorian physicists, the moral allegorization of thermodynamic laws represented a vigorous reactionary attempt to constrain scientific discourse within a Christian hermeneutics.[40] This energetic opposition to secular materialism was clearly the case with the popularization of thermodynamics in Tait and Stewart's *The Unseen Universe,* which operated on the heat-death scenario by arguing for the scientific plausibility of certain doctrines of personal immortality and constructing a grandiose allegorical cosmos that conflated the material world or "seen universe" with a spiritual realm or "unseen universe." *The Unseen Universe* bound its two-tiered world together with the flux of energies that radiate through the universal medium of the ether, the same "doubtful . . . ether" Lawrence remarked in 1925 as representing a scientific entity of dubious or at least unproved reality.

Ironically, Tait and Stewart's influential efforts to fit thermodynamic mechanisms within Christian revelation seem to have given considerable scientistic impetus to alternative spiritual systems. By ostensibly marshaling scientific credence for the divine dispensation of immaterial energies, this work strongly abetted the mythopoetic fervor of the late-Victorian and early-modernist eras. The cultural wake of the allegory of thermodynamics codified in *The Unseen Universe* impelled a series of visionary scientisms that lapped upon D. H. Lawrence's shores, from the theosophy of Madame Blavatsky and Annie Besant, to the monist energetics of Ernst Haeckel and Wilhelm Ostwald, to the hyperspace philosophies of the fourth dimension initiated by Charles Howard Hinton and later disseminated to Lawrence by the English translation of Ouspensky's *Tertium Organum*. That the spiritual figures of *The Unseen Universe* conceivably had some direct impact on Lawrence is suggested by his entrenched habit of valorizing the invisible, of equating the "unseen" with the essential.[41]

Lawrence was thus the recipient of a rather amorphous and contradictory inheritance of highly charged energic scientisms. For instance, in the continuation of the same passage from *The Crown* in which Lawrence inveighed against the "absoluteness" of the first law of thermodynamics, he nonetheless adapted the second law, as if by ingrained rhetorical reflex, to the range of unhealthy behaviors he called "sensationalism." "To get a sensation," Lawrence explained, "you reduce down some part of your complex psyche, physical and psychic. You get a flash, like when you strike a match. But a match once struck can never be struck again. It is finished.—Sensationalism is an exhaustive process."[42] These scientized notions of psychological and physiological entropy—the moral dissipation and irrecoverable waste of vital energies—are so pervasive in his writings that they could be handily labeled as instances of *D. H. Lawrence's second law.*

Fantasia of the Unconscious

When assessing his various doctrinal commentaries on science in general and the sun in particular, it is useful to recall that Lawrence came of age exactly during the heyday of heat death. Before and after the decade of the Great War, the science in which he had previously been schooled was in profound crisis. Along with so much else, its stable foundations, established verities, and universal continuities were being swept away. Like *The Crown, Fantasia of the Unconscious* is one of a number of discursive texts Lawrence wrote and rewrote in his persistent efforts to work out a deliberate system of existential ethics from the more intuitional composition of his poems and novels. This strange and often disturbing book is the full-length continuation of a shorter text entitled *Psychoanalysis and the Unconscious,* which had the immediate agenda of arguing with Freud's particular constructions of the psychological field. Lawrence's treatises on the unconscious set out a cluster of psychophysiological arguments intended to recenter familial and erotic relations on a complex electromagnetic interplay of nerve plexuses and sympathetic ganglions. Released in the United States in May 1921, *Psychoanalysis and the Unconscious* dumbfounded most of its first readers and reviewers. The finished form of the *Fantasia* represents Lawrence's defiant rehearsal and further elaboration of the same material.

The material itself is very much an acquired taste. We can get some of the flavor of his works on the unconscious from a wonderfully irreverent letter that Leo Stein, Gertrude's brother, actually sent to Lawrence, which Lawrence then passed on to Mabel Dodge Luhan, as it was her copy of *Psychoanalysis and the Unconscious* that Stein had read as her houseguest in Taos, New Mexico. Stein's perspicacious critique offers an explanation of the resistance Lawrence's electrified nerve-maps encountered from his prior journalistic outlets, and also refers Lawrence's framework of ideas to a tradition grounded in the philosophical biology developed in German *Naturphilosophie* in the early nineteenth century and extended in the work of Schopenhauer:

> Your work is of course in several respects in a familiar tradition. The title recalls Schopenhauer, and might well have been, The Fourfold Root of Sufficient Being. On the other hand it continues the work of the Natur-Philosophen. It makes me think of Oken and of science like the following: "Die Mathematik ist auf das Nichts begründet, und entspringt mithin aus dem Nichts." "Der Sehnerv ist ein organisirter Lichtstrahl, das Hirn eine organisirte Sonne, das Auge eine organisirte farbensonne, Regenbogen."[43]

As Leo Stein implied, the scientistic substance of the *Fantasia* amounts to a freewheeling reworking of mesmerism and theosophy, which is to say, the quasi-defunct scientific grounds of Lawrence's psychological system were already infused with the ambient electromagnetic fascinations of the pre-Einsteinian period. To some extent, however, that obsolescence released Lawrence's text to be the *fantasia* he titled it. Lawrence's *Fantasia* is indeed a kind of dream text encoding his prophetic desires in the signs and symbols of a mystic electrovital cosmos that repaired the emotional and spiritual gaps of the positivist scientisms he was reared in at the turn of the twentieth century.

In any event, wonder and the unknown remained for Lawrence higher values than the acceptance of any secular consensus about reality. And so he continued to spin out a kind of funky body mysticism in which new and old sciences jostled against archaic and modern forms of spiritual allegory. In a projected self-description that measures a culturally specific period response, Lawrence referred the broad modernist plunge back into psychologistic, occultic, and Theosophic mythomania to the collapse of the scientific positivism of the previous era. "These myths now begin to hypnotize us again, our own impulse towards our own scientific way of understanding being almost spent."[44] However, one effect of Lawrence's intellectual disillusionment was the rather petulant avowal, "I do not believe one-fifth of what science can tell me about the sun" (181).

Confident that he had disabused himself of both a bankrupt Chapel religiosity and a spurious secular-mechanistic creed, and convinced of the cultural need to fill the universal vacuum left by the deaths of God and of Science, Lawrence proceeded in *Fantasia of the Unconscious* to rewrite the Book of Genesis. In what he disarmingly defined as a "sudden lurch into cosmology, or cosmogony," he composed a mythic model of energy by which to reanimate the living world and to confer vital meaning upon its creatures. Along with this revision of materialistic cosmogony, however, Lawrence also turned to gestures inherent in the discursive productions of both Romantic and Victorian natural theology. The semiparodic cosmogony of Lawrence's *Fantasia* begins:

> In the beginning was a living creature, its plasm quivering and its life-pulse throbbing. This little creature died, as little creatures always do. But not before it had had young ones. When the daddy creature died, it fell to pieces. And that was the beginning of the cosmos. . . .
>
> Which is my account of the Creation. And I mean by it, that Life is not and never was anything but living creatures. . . . Out of living creatures the material cosmos was made: out of the death of living creatures, when their little living bodies fell dead and fell asunder into all sorts of matter and forces and energies, sun, moon, stars, and worlds. So you got the universe. Where you got the living creature from, that first one, don't ask me. He was just there.[45]

Lawrence's vitalistic materialism here is an implicit attack on the abstract vitalism of mystic idealists like Edward Carpenter, for whom the life force was a metaphysical power, the immaterial anima of the universe.[46] But his insistence that "Life is not and *never was* anything but living creatures" is reminiscent of certain Victorian arguments against the theory of evolution. Some of Darwin's critics had objected, even if new species evolved from previous ones, out of what did the very first species evolve? Pointing to the abiding mystery of the primal origin of life—about which Darwinian biology had little publicly to say—was a standard gambit of both mystic vitalists and orthodox Christians.

Interestingly, an argument similar to Lawrence's occurs in 1875 in the midst of the Presbyterian physics of *The Unseen Universe*. The immediate context of Tait and Stewart's discussion is the need to discriminate between *energy*—which concept they have been loading up with heavy theological freight—and *life*. The concept of life was the more typical recipient of spiritual amplification, and it had frequently been smudged over into that of physical or mechanical energy. Tait and Stewart's effort to desynonymize energy and life points out the very power of this particular equivocation within modern scientistic systems, as in Lawrence's own energetic vitalism. So they announced another Victorian scientific "law" on a par with the laws of thermodynamics, the "law of Biogenesis":

> Life proceeds from life, or to speak more accurately, . . . a conditioned living thing proceeds only from a conditioned living thing. . . . If we assume the truth of this principle it appears to lead us directly to infer that life is not merely a species of energy, or a phenomenon of matter. For we have seen that the great characteristic of all energy is its transmutability—its Protean power of passing from one form to another. . . . Life, however, can only be produced from life, and this law would seem to be an indication that the solution of the mystery is not to be found by making life merely a species of energy.[47]

By some form of cultural inheritance, then, Lawrence's cosmological fantasia seems to have reiterated the "law of Biogenesis," a principle that had expressed an objective epistemological limit *within* Victorian science. In this work entitled a fantasia of the *unconscious,* his allegorical leaps beyond the present—ripping "the veil of the old vision across"[48]—led also to unconscious reversions to the past.

For better and worse, the *Fantasia* is impelled primarily by *reversal.* Since the mid–nineteenth century, the cosmology of classical thermodynamics had posited the flow of solar energy as the material origin and maintainer of life: the dying sun portended death to all living things on earth. The *Fantasia* reversed that entire scenario: "Instead of life being drawn from the sun," Lawrence asserts, the sun "is the emanation from life itself, that is, from all the living plants and creatures which nourish the sun."[49] Such defiant lurches into scientistic cosmology are not without visionary appeal as deliberate provocations. However, the reversal at the level of physics in these mythopoetic claims also codes the reactionary cast of Lawrence's current sexual and social politics. Whether applied to the body or to the cosmos, some of Lawrence's revisionary scientisms are ultimately stranded within the closed paradigm of a "circuit of polarity" that descends, by way of the late-Victorian vogue for electromagnetic fields and the early-Victorian fad of electrobiology, from the grid of German Romantic physiology.[50] As with many of his modernist peers, such idiosyncratic attempts to break out of the impasses of the late-classical scientific cosmos often placed one in the grip of an even less desirable scientific ideology. Again in line with Zamyatin, Lawrence's conceptual reliance on polarity models and their "reversibility" could be a speculative cul-de-sac reinstating the previous era's broad cultural impasse between vitalism and mechanism. For instance, in the sermon from Rawdon Lilly that concludes *Aaron's Rod,* composed concurrently with the writing of the *Fantasia,* the rhetoric of polar reversal is inscribed directly on the demand for a revitalized masculinism:

> "We must reverse the poles. The woman must now submit—but deeply, deeply, and richly! No subservience. None of that. No slavery. A deep, unfathomable free submission."
>
> "You'll never get it," said Aaron.[51]

With regard to pronouncements such as Lilly's, Lawrence's novels provide a variable fund of dialogical irony, as in Aaron's perspicuous if downhearted retort. But in the monological declamations of the *Fantasia,* Lawrence eventually elaborates his cosmological polarity-rhetoric into a sexual puritanism as distasteful as the moral puritanism he otherwise despised: "The great thing is to keep the sexes pure. . . . pure maleness in a man, pure femaleness in a women. . . . And though you reverse the sexual polarity, the flow between the sexes, still the difference is the same."[52] Lawrence's reversions from Victorian to Romantic scientism functioned in such instances to foreclose the forms of open intercourse he otherwise wanted to celebrate.

Lady Chatterley's Lover

To some extent *Lady Chatterley's Lover* resolves the thermodynamic and evolutionary themes in Lawrence's later writing.[53] For most of the novel Connie Chatterley's sympathies play back and forth between hope and despair over imagined forms of evolutionary and devolutionary possibility. This thematic and emotional rhythm reaches its climax at the beginning of chapter 16, in the last great argument between Clifford and Connie, after she returns from her erotic streak through the rainstorm and flowery communion with Mellors, in anticipation of her departure for Venice. Clifford reads from "one of the latest scientific-religious books" a disquisition on evolution that asserts a cosmic inverse ratio inimical to the passional progress that the novel up to this point has been detailing:

> "What do you think of this, by the way?" he said, reaching for his book. "You'd have no need to cool your ardent body by running out in the rain, if only we had a few more aeons of evolution behind us. Ah, here it is!—'The universe shows us two aspects: on one side it is physically wasting, on the other it is spiritually ascending.'" (264)

Lawrence's last novel offers a reprise of the modernist scientisms that still entangled evolution together with thermodynamics, and that used the dark promise of physical entropy to enforce a spiritual or transcendental solution to humanity's cosmic dilemma. Once again Lawrence rejected the Victorian moralization of solar physics, when Connie retorts to Clifford, "Physically wasting? . . . Do you think the sun is smaller than he used to be? He's not, to me."[54] Connie's simple negation sums up Lawrence's career in relation to the mechanistic physics his generation inherited from late-Victorian science. In intuitive defiance of what was in fact a faulty formulation of imminent entropic collapse, Lawrence and his characters continued to affirm the inexhaustibility of the "creative mystery." Connie throws the different sun of her alternative, vital cosmos back at Clifford and his intellectual mentors: "But now the body is coming really to life, it is really rising from the tomb. And it will be a lovely, lovely life in the lovely cosmos, the life of the human body" (266).

The manifest elimination of gender polarity from this formulation realigns it with the androgynous or Carpenterian register of progressivist vitalism. The ontological polarism of *Lady Chatterley's Lover* is reasserted, however, in the clash of its two gods, Clifford's white mental-ideal or philosophico-scientistic deity of transcendental evolutionism and Connie's dark daemon of the living body. When Clifford insists: "Believe me, whatever God there is is slowly eliminating the guts and alimentary system from the human being, to evolve a higher, more spiritual being," Connie replies, "Why should I believe you, Clifford, when I feel that whatever God there is has at last wakened up in my guts, as you call them, and is rippling so happily there, like dawn?"[55]

The novel concludes with the contrast between Clifford's infantile sterility and the ineffable vital fecundity rippling in Connie's bowels. Mellors's final credo combines evolutionary vitalism and revisionary thermodynamics in another figuration of a different sun—an organic flame that "ripples" in waves of sexual energy through the universal ether, yet burns from a source that lies beyond the material world. "I believe," Mellors declares, "in a higher mystery, that doesn't let even the crocus be blown out. . . . We fucked a flame into being. Even the flowers are fucked into being between the sun and the earth. But it's a delicate thing, and takes patience and the long pause."[56] In the allegory of thermodynamics that accompanies Lawrence's own gospel of revelations, "the long pause" represents an alternative apocalypse. Substituted for the dismal finality of heat death is the peacefulness of a posthuman cosmos in which a living sun cohabits the cosmos with a living earth freely evolving eon by eon, in the aftermath of the "bad time coming . . . death and destruction, for these industrial masses" (342).

PART 3
ALLEGORIES OF DIMENSIONALITY

Chapter 7
The Ether of Space

Seas of bright juice suffuse heaven.
—*Whitman*

When the laws of thermodynamics were first formulated in the 1850s and 1860s, the concept of physical energy arrived already attached, like a Siamese twin, to another scientific hypothesis positing the universal presence of a subtle ethereal medium. This hypothetical ether supported the propagation of radiation through space, just as the elasticity of the atmosphere conveys acoustic vibrations, or as a body of water transmits undulations that rise to visibility on its surface. Thermodynamics first encountered space as a series of mechanical ether media crossed through with thermal and optical waves and electrical and magnetic "lines of force." For the popularizers of Victorian science, light and electromagnetic radiation were to be understood as waves that rippled within a cosmic sea of ether.

As the medium for the propagation of radiant energies—the mode of energic and informatic connections—the luminiferous ether was positioned precisely as the daemonic or intermediating term. The ether was thus a virtual site of cultural allegoresis. It served as an ideological screen for the anticipatory or hallucinatory projection of the communication technologies that would in fact unfold from the radio waves Heinrich Hertz discovered by extrapolating from Maxwell's equations for the electromagnetic ether. In particular, the rarefaction of the ether to the point of absolute impalpability primed it for a parallel, parascientific career as a medium of subtle and penetrating vibrations emitted from and received by prophetic and sensitive psyches.

The discourse of the ether memorializes an instance when institutional science in Europe and the United States invested heavily in the substantial reality of a purely theoretical construction. As a speculative entity elevated by some into a real material element, the ether struck certain Victorian thinkers as an epistemological scandal.[1] But most specialists and the general public accepted the reality of the ether as scientific dogma. Eventually the mechanical conception of the ether as a spatially fixed and continuous substance was abandoned, and the luminiferous ether of late-classical physics was ultimately broken down and reassembled in the paradigm shift from Newtonian dynamics to quantum mechanics.[2] Einstein's special theory of relativity announced the ether's conceptual devolution as a scientific object in 1905, but with some assistance from Einstein himself the ether has persisted in the shadows of relativity theory. Recently the ether has been revisited in the discussion of quantum field theory.[3]

Throughout the era of classical thermodynamics, at least, the concept of the ether maintained its cultural power as a medium for energetic allegory. Much of the imaginative ambiance of modern space descends to us from this prior, ethereal form. Developed over centuries as a reputable scientific theory, the ether medium also became a definitive object of visionary scientism. Even while the scientific status of the ether was being challenged, as a transformative and connective medium rippling with undetermined energies it continued to inflect the visionary forms of modern space-concepts. Subsumed by modern field theory, the ether's nineteenth-century thermodynamic and electromagnetic imprints remain culturally embedded in the four-dimensional constructions and scientistic space-fictions that began to emerge in Victorian and modernist literature and art. The doctrine of the ether was a conceptual preparation for the theory of general relativity and the technologies of wireless radio and television transmissions. But it was also a profound occasion for visionary scientisms within the cultural imaginary, providing a textual if not a substantial medium between material sciences, spiritual systems, and science fictions.

The Ether Solution

Before the twentieth century, the phenomena of radiant energies such as light, heat, and electromagnetism were often explained as the propagation of wave motions within the medium of a universal ethereal substance. Newton also postulated an ether to account for the "action at a distance" of gravitation.[4] One or more subtle substances were thought to fill the void of space, providing a means of mechanical connection among the various material bodies distributed in the cosmos.[5] In its earlier development from the pre-Socratics to the physical models of Descartes, Newton, Huygens, Young, and Fresnel, the ether was one of a number of imponderable fluids postulated to account for the dynamics of invisible and radiant phenomena. For an *Encyclopedia Britannica* article, Clerk Maxwell sketched the course of ether theories in the centuries before the consolidation of the energy concept abolished most of them from physical theory:

> To Descartes, who made extension the sole essential property of matter, and matter a necessary condition of extension, the bare existence of bodies apparently at a distance was a proof of the existence of a continuous medium between them. But besides these high metaphysical necessities for a medium, there were more mundane uses to be fulfilled by aethers. Aethers were invented for the planets to swim in, to constitute electric atmospheres and magnetic effluvia, to convey sensations from one part of our bodies to another, and so on, till all space had been filled three or four times over with aethers.[6]

The theory of the ether may be compared to the concept of *caloric,* another subtle substance of earlier physics. In the immediate prehistory of thermodynamics, the movement of heat was conceived as an imponderable flow of caloric into and out of grosser substances. The conservation of caloric, its nondegradation throughout the working cycle—the conceptual scaffolding for Sadi Carnot's ideal heat-engine—preceded the mid-nineteenth-century reformulation of physical forces in terms of the conservation of energy.[7] The modern form of "Carnot's principle," the second law of thermodynamics, emerged when the presumed materiality of caloric sublimated into the kinetic energies of molecules. The conservation of caloric was discarded in favor of an energetics focused on energy and entropy—the temperature differentials and transformations necessary for the mechanical transformation of heat into work. But that successful elaboration of thermodynamic theory rested on what proved to be a scientific fiction, the "hypostatic fluid" of caloric.[8]

During the main period of its scientific viability, the ether concept accommodated and to some degree replaced the notion of absolute space in Newtonian mechanics. As the luminiferous ether posited by the wave theory of light gained credence in the nineteenth century, it tended to be coordinated with the absolute parameters of Newtonian space, and thus to be conceived as a stationary medium through which all other material bodies were in relative motion. But in the nineteenth century the fabric of absolute Newtonian space wore increasingly thin. Donald Benson has listed three "anomalies" between Newtonian and mid-nineteenth-century theory of major concern. First, by midcentury the wave theory of light had fully overturned Newton's corpuscular model. There were also growing uncertainties as to the ultimate solidity of matter, as well as the suspicion that matter, like light, might also consist of motions within a medium rather than fixed particles. Finally, non-Euclidean geometries had begun to challenge the notion of an absolute and finite set of spatial coordinates. In this situation, the concept of a universal ether—a stationary cosmic medium for the propagation and persistence of vibratory motions, through which stars, planets, atoms, and molecules wheeled—provided the "means for resolving the anomalies which absolute space could not and for saving the continuities they required."[9]

The mechanical ether-concept provided at best a paradoxical resolution to the theoretical difficulties created by extending a material medium upon the Newtonian void. The Victorian ether encapsulated the conceptual binds generated when a physics of absolute mechanism confronted the relativistic phenomena of radiant energies. For instance, it was difficult to imagine what state or phase this ethereal matter was in. It was subtler than any known gas, it could not be evacuated from a vacuum, and yet by transmitting light as transverse rather than longitudinal waves, it acted like a crystalline solid. If the ether were a fluid, the movement of bodies through it should create some sort of turbulence, eddies disrupting the waves it conveyed. But no such disturbances could be noted. It was postulated instead that the ether was so rarefied that gross matter passed through it without friction. At the beginning of the nineteenth century, the resuscitator of Huygens's wave theory of light, Thomas Young, supposed that "the luminiferous ether pervades the substance of all material bodies with little or no resistance, as freely perhaps as the wind passes through a grove of trees."[10]

As the physical branches of thermodynamics and electromagnetism were being consolidated, then, ether theory persisted in a state of fluctuation. With the unification and interconvertibility of energies codified by the first law of thermodynamics, by midcentury only two ethers remained in mainstream physics—the medium presumed to carry electrical and magnetic forces, and the luminiferous ether credited with the propagation of light. We noted in chapter 4 how Maxwell constructed theoretical grounds for the unification of these two ethers. As a result, by the 1870s the only ether seriously at play in physical theory was the luminiferous/electromagnetic medium of late-classical wave theory. Instead of several ethers conveying particular forms of energy, now a single ether conveyed multiple energies.

By this time careful measurements of reflective and refractive phenomena had suggested a notion of "ether drag," in which certain substances were conceived to absorb a small portion of the ether through which they passed. Maxwell assessed this aspect of the question in his ether article. With regard to the possibility of these relative motions of the ether, "we must therefore consider the aether within dense bodies as somewhat loosely connected with the dense bodies, and we have next to inquire whether, when these dense bodies are in motion through the great ocean of aether, they carry along with them the aether they contain, or whether the aether passes through them as the water of the sea passes through the meshes of a net when it is towed along by a boat."[11]

Unlike some of his more enthusiastic colleagues, Maxwell was forthright about the difficulties that beset the theory of the ether, especially when defined as a form of elastic solid. He was also sensitive to the dissipative aspects of the ether problem. As a material medium through which energies were propagated as transverse displacements of its substance, friction ought eventually to occur, by which those energies would be "dissipated into that irregular agitation of the medium which, in ordinary media, is called heat."[12] And yet across cosmic lengths of space, light and electromagnetic waves never appeared to die out. Although Maxwell admitted the inability of science to conceive the actual inner mechanism of the ether that supported such phenomena, he nonetheless affirmed its reality. "Whatever difficulties we may have in forming a consistent idea of the constitution of the aether, there can be no doubt that the interplanetary and interstellar spaces are not empty, but are occupied by a material substance or body, which is certainly the largest, and probably the most uniform body of which we have any knowledge" (2:775).

On the authority of such affirmations from the most eminent of theoretical physicists, the quest for the material reality of the ether continued with quickened pace in the last quarter of the nineteenth century. The successful construction of sophisticated apparatuses to determine the velocity of light through various media—air, water, glass—encouraged experimenters to devise means for the empirical confirmation and analysis of the ether. Without success: the famous Michelson-Morley investigations of the 1880s produced a definitive null result.[13] The need to account for this nonfinding was a crucial step toward the reorientation of physics away from classical absolutes and invariances and toward the relative and discontinuous cosmos we now inhabit.[14] But what especially concerns us in this study is the development parallel to the scientific career of the ether—its cultural migration from authorized explanatory concept within late-classical mechanics to a key component in a series of scientistic cosmologies spun off from Victorian physics. The technoscientisms of the ether generated a wide-ranging cultural poesis of energy in space. When physical speculation reached the boundary of experimental experience, the ether model provided the bridge over which one might proceed to metaphysical matters while ostensibly maintaining a materialistic grounding.

John Tyndall's Imagination

Maxwell's work represents the high point of the heuristic application of the ether concept. A more or less plausible form of the ether as an electromagnetic medium was actively constructed as part of a successful theoretical investigation of physical phenomena. However, Maxwell's nuanced and mathematized approach was uncommon. Rhetorical inflation was more often the case, and scientistic poesis was directed more often at the ether's presumed functions than at its presumed structure. In this regard, no scientist of the Victorian period responded more rhapsodically to the conceptual resolutions provided by the ethereal medium than John Tyndall, "probably the most important single figure among the scientific community for those concerned with the relations between ordinary human perceptions and science."[15] The following passages are taken from his two-volume *Fragments of Science* and his text on thermodynamics, *Heat: A Mode of Motion,* both of which collected and revised his various professional addresses to specialized and general audiences. I will quote them at some length to convey the tenor of Tyndall's eloquence.

Tyndall asserted the latent existence of the ether on the basis of a complex of manifest effects, otherwise inexplicable, that were mutually comprehensible as wave phenomena. His faith in the substantial reality of the ether also seems connected to his form of spiritualized materialism. Indeed, "despite denying vitalism—and indeed all religious authority—he contrived to keep some dark place for an ordering, though not a pre-emptive, Intelligence."[16] Typically for this agnostic Irish Protestant and disciple of Carlyle, Tyndall clothed the ether in the accents of the natural sublime:

> The notion of this medium must not be considered as a vague or fanciful conception on the part of scientific men. Of its reality most of them are as convinced as they are of the existence of the sun and moon. The luminiferous ether has definite mechanical properties. It is almost infinitely more attenuated than any known gas, but its properties are those of a solid rather than of a gas. It resembles jelly rather than air. . . . In fact it is the vehicle of [star] light, and without it they could not be seen. This all-pervading substance takes up their molecular tremors, and conveys them with inconceivable rapidity to our organs of vision. It is the transported shiver of bodies countless millions of miles distant, which translates itself in human consciousness into the splendour of the firmament at night.[17]

Tyndall encouraged his general audiences to conceive of the effectivity of the ether by exercising their powers of mechanical visualization. He posited the ether as a thermal as well as a luminiferous medium, as the vehicle of radiant heat as well as light across both mundane and cosmic distances: "You must . . . figure the molecules of the heated body as in a state of motion; you must figure that motion as communicated to the surrounding ether, and transmitted through it with a velocity which we have the strongest reason for believing to be the same as that of light. When, therefore, you turn towards a fire on a cold day and expose your chilled hands to its influence, the warmth which you feel is due to the impact of these ethereal billows upon your skin."[18] Tyndall credited the all-accommodating nature of the ethereal matrix with the transmission of both light and heat, supporting an infinite mixture of energy vibrations, ultimately creating a symphonic concordance of discrete waves. Tyndall's musical metaphors for ethereal effects are explicitly extended from the tangible realm of sound waves in motion through air:

> The ether is never at rest, for in the absence of light-waves we have heat-waves always speeding through it. In the spaces of the universe both classes of undulations incessantly commingle. Here the waves issuing from uncounted centres cross, coincide, oppose, and pass through each other, without confusion or ultimate extinction. Every star is seen across the entanglement of wave-motions produced by all other stars. It is the ceaseless thrill caused by those distant orbs collectively in the ether, that constitutes what we call the "temperature of space." As the air of a room accommodates itself to the requirements of an orchestra, transmitting each vibration of every pipe and string, so does the inter-stellar ether accommodate itself to the requirement of light and heat.[19]

Operating in tandem with thermodynamic paradigms, Tyndall's theory of "calorescence" models the ethereal transmission of solar and mundane heat on the "luminescent" transmission of light waves. For Tyndall the first law of thermodynamics determines a perpetual recycling of ethereal vibrations, a continuous universal unity of energies in cosmic harmony: "And whether they break in foam upon the shore, or rub silently against the ocean's bed, or subside by the mutual friction of their own parts, the sea waves, which cannot subside without producing heat, finally resolve themselves into waves of ether, thus regenerating the motion from which their temporary existence was derived. . . . Thus is sentient man acted on by Nature, the optic, the auditory, and other nerves of the human body being so many strings differently tuned, and responsive to different forms of the universal power."[20]

The cultural accommodations offered by the ether concept are virtually evident in Tyndall's descriptions of the intellectual satisfactions, the aesthetic and cognitive delights, it affords: "By the assumption of this wonderful intangible *ether,* all the phenomena of optics are accounted for, with a fullness, and clearness, and conclusiveness which leaves no desire of the intellect unsatisfied."[21] However, despite his claim that an ether filling space was as factual as the force of gravity itself, Tyndall's rhetoric began to tip in the direction of its own fictionality when he started to reflect on his present methodology as an instance of imaginative extension and heuristic visualization:

> If a single phenomenon could be pointed out which the ether is proved incompetent to explain, we should have to give it up; but no such phenomenon has ever been pointed out. It is, therefore, at least as certain that space is filled with a medium by means of which suns and stars diffuse their radiant power, as that it is traversed by that force [gravity] which holds in its grasp, not only our planetary system, but the immeasurable heavens themselves.
>
> There is no more wonderful instance than this of the production of a line of thought, from the world of the senses into the region of pure imagination. I mean by imagination here, not that play of fancy which can give to airy nothings a local habitation and a name, but that power which enables the mind to conceive realities which lie beyond the range of the senses—to present to itself distinct images of processes which, though mighty in the aggregate beyond all conception, are so minute individually as to elude all observation.[22]

In one of Tyndall's most celebrated lectures, "The Scientific Use of the Imagination," an address to the 1870 meeting of the British Association for the Advancement of Science, he defended a liberal use of imaginative projection in scientific investigation against the "Tories" of science "who regard Imagination as a faculty to be feared and avoided rather than employed."[23] Against this epistemological fastidiousness Tyndall urged the examples of a series of great British innovators—Newton, Dalton, Davy, and Faraday—whose successful theories resulted from bold hypothetical strokes. The centerpiece of Tyndall's arguments in favor of a liberal scientific imagination was—the ether hypothesis. In other words, Tyndall's science as well as his popular public lectures were allegorical performances. The luminiferous ether in which Tyndall placed his scientific conviction is precisely and by his own account the result of a poetic act, one that endows all space with an analogical substance conceptually extended from the "aerial medium" of sound:

> We know intimately whereon the velocity of sound depends. When we lessen the density of the aerial medium, and preserve its elasticity constant, we augment the velocity. When we heighten the elasticity, and keep the density constant, we also augment the velocity. A small density, therefore, and great elasticity, are the two things necessary to rapid propagation. Now light is

> known to move with the astounding velocity of 186,000 miles a second. How is such a velocity to be obtained? By boldly diffusing in space a medium of the requisite tenuity and elasticity. (2:106)

Tyndall's passage is striking in the way that it mingles multiply confirmed experimental results—the velocities of sound and light—with the heuristic panache to hypostasize a material ether, despite its severely paradoxical qualities, as a necessary conclusion from observable phenomena. The "boldness" at stake here was not Nature's, so to speak, in devising such an ultimate substance, but Science's—Science as manifested by intrepid discoverers willing to boldly go beyond the limits of empirical evidence.

Tyndall had good grounds for crediting his analogical procedure with scientific reliability. As we have seen, Maxwell had recently developed his theory of the electromagnetic field, "On Physical Lines of Force"—positing the unity of the electromagnetic and luminiferous media and thereby strengthening the theory of a single universal ether—on the basis of certain physical analogies suggested by Faraday and Thomson between electricity, heat, and elastic media. But whereas Maxwell was famously reticent about reifying his functional analogies, preferring to preserve their fictional or heuristic status, Tyndall favored the bolder literalizing gesture. His confidence in the material existence and the imminent demonstrability of the ether was such that he rested his further conviction in the substantial reality of atoms—which were also as yet undemonstrated hypothetical entities—*on* his belief in the ether.

Tyndall's scientific poetics, then, contain a typical sort of historical irony. The subsequent unfolding of the composition of the primary atomic particles would confirm to some degree his materialist convictions. At the same time, as his particular version of the ether—the ultimate matrix of those convictions and the last bastion of classical mechanism with its spatial absolutes and material continuities—was dismantled as a working scientific hypothesis, many of Tyndall's methodological exhortations were unraveled in the process. In the last decades of the nineteenth century, however, with some impetus from Tyndall's enthusiastic popularizations, the ether was gradually reincarnated as a cultural artifact, as less experimentally grounded speculators invested their conviction in analogical productions based on scientific allegories. Riding on the crest of the credence granted it by late-classical physics, the ether concept was eventually diffused into other cosmological discourses. From Tyndall's eloquent positivist scientism of an ether attuned to the vibrations of the human imagination, it was transposed to further zones of active visionary scientism.

The Ether and the Fourth Dimension

The Unseen Universe melded Victorian physics and metaphysics. Its well-established and socially conservative authors offered a theologized cosmology with the polemical intention of coordinating the new laws of energy with the letter of Christian revelation. For them, the very physical mysteriousness of the luminiferous ether intimated that this medium functioned within an ultimate cosmos divided between mundane and divine realities: "Regard it as we please, there can be no doubt that the properties of the ether are of a much higher order in the arcana of nature than those of tangible matter."[24]

The thermodynamic allegory of divine energy in Tait and Stewart rests precisely on the luminiferous ether as the paraspace or intermediating zone that connects known space to the "unseen universe": "What, then, is the Unseen to which they raise their thoughts?" Maxwell asked in his gentle review of Tait and Stewart's follow-up volume, *Paradoxical Philosophy:* "In the first place the luminiferous ether, the tremors of which are the dynamical equivalent of all the energy which has been lost by radiation from the various systems of grosser matter which it surrounds."[25] With that wonderfully pliable hypothetical medium of Victorian science Tait and Stewart sutured their cosmic totality of seen and unseen worlds together. The speculative conclusion of the *Unseen Universe* offers a lyrical intimation of the ether's vast capacity, in particular, to overcome the universal effects of thermodynamic dissipation by opening up a space of energic recuperation. To initiate this line of thought, the authors considered the negligible size and solar requirements of the earth relative to the sun's total radiation of energy in all directions. About the sun's "wasting" of its energy it is asked: "Can anything be more perplexing than this seemingly frightful expenditure of the very life and essence of the system?"[26] But it is conceivable, they continued, that by means of the ether the invisible world retrieves the energy wasted in the visible:

> Now, is it not natural to imagine that a universe of this [unseen] nature, which we have reason to think exists, and is connected by bonds of energy with the visible universe, is also capable of receiving energy from it? Whether is it more likely that by far the larger portion of the high-class energy of the present universe is traveling outwards into space with an immense velocity, or that it is gradually transferred into an invisible order of things? May we not regard ether or the medium as not merely a bridge between one portion of the visible universe and another, but also as a bridge between one order of things and another, forming as it were a species of cement, in virtue of which the various orders of the universe are welded together and made into one?[27]

As we noted in chapter 4 with regard to the deployment of Maxwell's Demon, the *Unseen Universe* carries out an allegorical defense against thermodynamics itself—one that also extends to the ether concept.[28] Their phantasmagoric cosmos defends against the temporal destiny of energic dissipation by recourse to the hypothetically nondissipative medium of solar transmissions through space. In this way space trumps time—the entropy of the material world as a whole is displaced in favor of a transformative and recuperative ether. The human soul is reprieved from its death sentence by a paraspatial afterlife.

Just as "it is not altogether surprising that spiritualists, whom he detested, sought to claim Tyndall as a medium,"[29] Tait and Stewart's scientistic defense of cosmic immortality also inspired the vigorous spiritualistic movements of its day. Two years after the appearance of *The Unseen Universe,* in 1877 Madame Blavatsky published *Isis Unveiled: A Master-Key to the Mysteries of Ancient and Modern Science and Theology.*[30] Here she cited *The Unseen Universe* at length to lend scientific prestige to her eclectic cosmos, typically using Tait and Stewart's Christianized physics to contribute to her scientized allegoreses of world mythology.[31] Madame Blavatsky also cited the new science of *The Unseen Universe* in support of a metaphysical ether well known to the magicians and miracle workers "of all periods, schools, and countries."[32] Chapter 5 of *Isis Unveiled* was devoted to "The Ether, or 'Astral Light.'" For Blavatsky the luminiferous ether of late-classical physics brought scientific confirmation to the occult lore of previous ages. The initiates of this wisdom "produced their wonders, because they were perfectly familiar with the imponderable—in their effects—but otherwise perfectly tangible waves of the astral light. They controlled the currents by guiding them with their will-power."[33]

Blavatsky was not especially interested in *The Unseen Universe*'s thermodynamic apocalypse, nor did she seem to cognize entropy or dissipation per se. Rather, she appropriated the recuperative side of Tait and Stewart's arguments at the point that the invisible world recovers the spilt energy of the material world and so stocks its celestial coffers with eternal potency. But what the British physicists and the Russian émigré theosophist shared was a total investment in the hyperreality of the ether: "What is the primordial Chaos," Blavatsky asks, "but Aether? . . . Ether, the celestial virgin, the spiritual mother of every existing form and being, from whose bosom as soon as 'incubated' by the Divine Spirit, are called into existence Matter and Life, Force and Action."[34]

In *The Unseen Universe,* ether as thermodynamic space provided a zone of energy transfer that countered the irreversibility of thermodynamic time.

In *Isis Unveiled,* the technoscientistic ether was further conceived as a kind of cosmic data bank, and the daemonic zone of ethereal intercommunication crossed over from an energic to an informational medium. As in *The Coming Race* of Bulwer-Lytton, one of Madame Blavatsky's favorite novelists,[35] we have a kind of cybernetic premonition. The Victorian retooling of entropy as theological allegory—that is, as a divine writing of cosmic significance—anticipates the postmodern retooling of entropy as a universal measure of information. All of these works of the 1870s presented the luminiferous ether as counterentropic, specifically, as a medium where temporal energy becomes paraspatial information. Tait and Stewart framed this idea with something like the "future light cone" of contemporary cosmology.[36] They speculated that the seemingly wasted energy of the sun and other stars "may only be an arrangement in virtue of which our universe keeps up a memory of the past at the expense of the present. . . . A picture of the sun may be said to be traveling through space with an inconceivable velocity, and, in fact, continual photographs of all occurrences are thus produced and retained. A large portion of the energy of the universe may thus be said to be invested in such pictures."[37] In *Isis Unveiled,* the *Unseen Universe*'s vision of a cosmic picture archive, with its scenario of wasted energy recouped within the ether as information, was immediately seized on to credit her occult mythopoeses with scientific authority:

> Ancient as well as modern wisdom, vaticination and science, agree in corroborating the claims of the kabalists. It is on the indestructible tablets of the astral light that is stamped the impression of every thought we think, and every act we perform . . . . The impressions of scenes and landscapes which the astral body saw during its peregrinations [during sleep] are still there, though lying latent under the pressure of matter. They may be awakened at any moment, and then, during such flashes of man's inner memory, there is an instantaneous interchange of energies between the visible and invisible universes. Between the "micrographs" of the cerebral ganglia and the photo-scenographic galleries of the astral light, a current is established.[38]

The energic ether-bridge set forth in the *Unseen Universe* became in theosophy a technology of thought-transmission, by which we see that Thomas Pynchon's Nefastis Machine was sired by James Clerk Maxwell upon Madame Blavatsky. In these florid theistic texts of the 1870s, the cultural discourse of thermodynamics sublimated a theory of energy-transformations into an ethereal ideology of information in perpetual cosmic circulation. The theosophical imagination in particular seems to have run on ahead of the development of global and galactic wireless electronic communications technology, and so to have prepared in some degree for further spiritualizations of energy in modernist aesthetics.

The *Unseen Universe* and *Isis Unveiled* demonstrate that by the end of the 1870s, the spatiality constituted by the universal medium of the ether is precisely thermodynamic space, an energic apprehension of space revised through the transformative and dissipative medium of thermodynamic time. The datedness or Victorian specificity of the ether as a paraspace would lie in its manifest unity, its structural function as an operator of continuity within an ostensibly coherent Christian or theosophical cosmos, relative to the dispersed reference-frames of Einsteinian relativity or the chaotic topoi of contemporary theory and fiction. But the protomodernity of these texts lies in their virtual anticipation of the line of thermodynamic development, from physics to cybernetics, from energy to information. The phantasmagoric spaces of the Victorian ether formed a significant part of the allegorical matrix out of which the modern paraspaces of aesthetic, pictorial, and science-fictional extrapolation have emerged. Victorian allegories of energy primed space for the uncanny transformations of its modern cosmological narrations.

In the 1870s and 1880s the cultural discourse of the luminiferous ether joined up with another influential scientistic theory of transformed space, the fourth dimension. The fourth dimension of space, originally based on mathematical rather than physical considerations, borrowed some of the glamour of electromagnetic effectivity and compounded the ether's apparent qualities of abolishing material limitations. Both conceptual artifacts bid for experimental verification; both were ultimately displaced and reformulated by Einstein's theories of relativity. Yet they are something more than just disciplinary anachronisms or phantasmagoric fictions of Victorian science. What Einstein abolished was the postulate that the ether formed an absolute frame of spatial reference on the model of Newton's absolute space. Nevertheless, the concept of relativity does not eliminate the energic effectivity of space itself as a medium for propagations, as something more than a mere void between material objects. According to physicist Frank Wilczek, this aspect of the ether idea was simply absorbed into and renamed as field theory. Similarly, Charles Howard Hinton's philosophy of hyperspace still commands attention because of its influence on the discourse of a series of modernist artistic movements, which in turn have inspired the architects of postmodern virtuality.[39] Just as the ether leaves its traces on field theory and the Ethernet, hyperspace lives on in the realms of cyberspace.

In the fifth edition of the *Unseen Universe,* published in April 1876, P. G. Tait and Balfour Stewart briefly adapted their paraspatial ether scenarios to a new idea of space recently in vogue. In the midst of a section devoted to embodying their ideas of the unseen in concrete form, they interpolated the following passage: "Just as points are the terminations of lines, lines the boundaries of surfaces, and surfaces the boundaries of portions of space of three dimensions:—so we may suppose our (essentially three-dimensional) matter to be the mere skin or boundary of an Unseen whose matter has *four* dimensions. And, just as there is a peculiar molecular difference between the surface-film and the rest of a mass of liquid—wherever such a surface-film exists, even in the smallest air-bubble—so the matter of our present universe may be regarded as produced by mere rents or cracks in that of the Unseen."[40]

The discourses of electromagnetism and thermodynamics popularized in the previous generation by scientists such as Faraday, Maxwell, Thomson, and Tyndall also provided Hinton with an authorized conceptual matrix for the fourth dimension of space. The immediate scientific (as opposed to mathematical) prototype of Hinton's hyperspace was the postulated but unverified luminiferous ether. Hinton's 1884 paper "What is the Fourth Dimension?" alludes to the ether hypotheses of late-classical wave theory in the phrase "ultimate particles": "Our proportions in [the fourth dimension] must be infinitely minute, or we should be conscious of them. If such be the case, it would probably be in the ultimate particles of matter, that we should discover the fourth dimension, for in the ultimate particles the sides in the three dimensions are very minute, and the magnitude in all four dimensions would be comparable."[41] For instance, in "The Constitution of Nature," Tyndall stated: "Two planets clash and come to rest; their motion, considered as that of masses, is destroyed, but it is in great part continued as a motion of their ultimate particles. It is this latter motion, taken up [by] the ether, and propagated through it with a velocity of 186,000 miles a second, that comes to us as the light and heat of suns and stars."[42] Hinton located in the "ultimate particles of matter" the threshold between the third and the fourth dimensions, but beyond that ethereal threshold he also located an immaterial "ultimate medium" that abrogated time by providing instantaneous connectivity to every point of three-dimensional space—the spatial fourth dimension proper.

Hinton argued that the hypothesis of a fourth dimension of space could alleviate some of the theoretical difficulties sustained by the mechanical ether concept, as well as explain its evident capacities to abolish or suspend spatiotemporal limitations in the world of three dimensions. Following directly from *The Unseen Universe*'s discussion of surface tensions, Hinton remarks: "In the surface of a fluid different laws obtain from those which hold throughout the mass. . . . It may well be that the laws of our universe are the surface tensions of a higher universe."[43] For Hinton, then, the ether is not itself the fourth dimension, but the medium between dimensions, the hypersurface that forms at the boundary between the three-dimensional world and the fourth dimension of space. The fourth dimension may be extended only to an infinitesimal distance beyond the ether-surface that forms when the three-dimensional universe rests upon the fourth. With respect to scientific plausibility, this theory fits the fourth dimension to the acceptable paradoxicality of the ether concept, by confining the fourth dimension's depth to what we would now call a "quantum level" of spatial extension: "We will suppose, then, that we are not in, but on the aether, only not on it in any known direction, but that the new direction is that which comes in. The aether is a smooth body, along which we slide, being distant from it at every point about the thickness of an atom; or, if we take our mean distance, being distant from it by half the thickness of an atom measured in this new direction" (57). Due to the near-depthlessness of this four-dimensional depth, the inverse-square law for calculating gravitational attraction did not yield to an inverse-cube law. Hyperspace would preserve even while suspending the laws of physics.[44]

Hinton constructs this theory of an ultrathin fourth dimension by figurative extension from the imaginary geometry of a two-dimensional plane world that rests upon an elastic-solid ether-surrogate, a frictionless crystalline bubble. By this non-Euclidean maneuver Hinton essentially merged the ether and the fourth dimension into a composite structure supplementing three-dimensional space with an infinitely elastic, hypersolid 4-D tissue: "It is so thin that over distances which we can measure the influence proceeding from a body is not lost by spreading in this new kind of depth."[45] Hinton develops this scenario such that energies within the ether—the wave motions of the ultimate particles—vibrate in the fourth dimension, the ultimate cosmological medium. Like the absolute matrix of the classical ether, then, the fourth dimension of space gathers up the universe, provides a foundation for all things, and produces the coherent continuity of the material world by mediating the totality of its particles:

> Thus the supporting body resembles, as far as we know it, a portion of a vast bubble. . . . The thin film—thin though infinitely extended in any way which we can measure—this thin film vibrates and quivers in this new direction [i.e., the spatial fourth dimension]. . . . The æther is more solid than the vastest mountain chains, yet thinner than a leaf; undestroyed by the fiercest

> heat of any furnace, for the heat of the furnace is but its shaking and quivering; bearing all the heavenly bodies on it, and conveying their influence to all regions of what we call space. (1:192–93)

Hinton's problem with this particular adaptation of the elastic-solid ether theory is a matter not so much of its practical validity as a scientific hypothesis as of its existential consequences. An ultrathin medium retains marginal physical plausibility at the price of foreclosing full four-dimensional extension, and thus the possibility of substantial four-dimensional hyperbeing. Hinton responds to this speculative challenge with a wonderful dramatization of the sanctioned Newtonian idea that the ether provides an absolute frame of spatial reference: "A being able to lay hold of the æther by any means would, unless he were instantly lost from amongst us by his staying still while the earth dashes on—he would be able to pass in any space direction in our world. . . . He would see all things differently."[46] To hold on to the possibility that one might substantially "lay hold of the æther," Hinton will make a breach in his thin-medium theory to allow for higher solidity within the fourth dimension of space.

Platonic Dimensions

As we have noted, in Platonic allegory the material world is only a "likeness," an iconic allusion to the realm of Forms. The world is a transient and turbulent emblem of eternal verities, and about the origin and constitution of this complex natural signifier the best that philosophy can offer is only, says Timaeus, a "likely story."[47] Platonic cosmology already constitutes the physical world as an allegorical structure originating from the temporal devolution of a timeless perfection. And culturally considered, in the nineteenth century Plato's realm of Forms was still the touchstone of Victorian idealists seeking rational authorization for an "unseen universe" or a "fourth dimension." However, Plato's influence on subsequent modes of allegorical construction extends beyond the mere bipartition of the cosmos into intersignifying realms of being and becoming.

The cosmology of the *Timaeus* also clarifies how the mathematical texture of Platonic rhetoric inflects Hinton's own four-dimensional constructions. The key technique exploited by Hinton and other mathematical allegorists is the device of continued geometric proportion. The *Timaeus* provides an incisive example of proportional analogy, for the specific purpose of binding the cosmos fabricated by the Demiurge into a unity, and notably, with specific reference to the possibility of alternative dimensions:

> Now that which comes to be must be bodily, and so visible and tangible; and nothing can be visible without fire, or tangible without something solid, and nothing is solid without earth. . . . But two things alone cannot be satisfactorily united without a third; for there must be some bond between them drawing them together. And of all bonds the best is that which makes itself and the terms it connects a unity in the fullest sense; and it is of the nature of a continued geometrical proportion to effect this most perfectly. . . .
>
> Now if it had been required that the body of the universe should be a plane surface with no depth, a single mean would have been enough to connect its companions and itself; but in fact the world was to be solid in form, and solids are always conjoined, not by one mean, but by two. Accordingly the god set water and air between fire and earth, and made them, so far as was possible, proportional to one another, so that as fire is to air, so is air to water, and as air is to water, so is water to earth, and thus he bound together the frame of a world visible and tangible. (31b–32b; trans. Cornford)

In his running exegesis of the dialogue, Francis Cornford explains that the Greek term for a continued geometrical proportion is exactly αναλογια, analogy. As in the parable of the cave, analogical series are the dynamos that generate Platonic allegories. Timaeus discusses the geometrical proportions that can be established among series of whole numbers, series of "planes" or mathematical squares—multiples of two numbers, as in the determination of two-dimensional areas—and series of "solids"—multiples of three numbers, as in the determination of three-dimensional volumes. Take for instance the continued geometric progression 2, 2^2, and 2^3: 2 is to 4 as 4 is to 8, and 8 is to 4 as 4 is to 2, and conversely, 4 is to 2 as 8 is to 4. However, a consideration of the three-dimensionality of material extension in the created world leads Plato to argue the logical necessity for there being *two* means of connection between the primal opposites fire and earth. Unlike squares, solids—measurable by multiplying the coordinates of all three dimensions—are connected in geometric series by two means. Therefore, both air and water must accompany fire and earth, for a three-dimensional world can be perfectly bound in union only by *four* elements.

I do not mean to suggest that Plato set out in the *Timaeus* a model of the fourth dimension of space, only that he provided the tools and techniques by which that construction could be produced by figurative extension from typical Platonic proportions. For Hinton's four-dimensional project is quite specifically a speculative effort of mathematical cosmology, produced by the application of proportional analogy to the interrelation of dimensional spaces. The dimensional ratios of hyperspace theorists are not isomorphic with the continued geometric proportions treated in the *Timaeus,* but they are, loosely speaking, sufficiently *analogous* to be extendable.

Hinton's Gramophone

At its conceptual inception, Hinton's hyperspace emerged within the turbulent cultural field set up between physical and metaphysical discourses of ultimate forces. By conveying the aura of the luminiferous ether to the allegorical machinery of Platonic geometry, Hinton's fourth dimension profoundly internalized modern divisions between empirical materialism and eclectic mysticism. That amalgam bifurcated Hinton's presentations into two incompatible models—one inserted into the discourse of Victorian physics, and another that gestured by way of ethical ideals to the scientistic spiritualism that spilled over from the nineteenth century into early modernism. The infinite connectivity of the ultimate medium accommodated the energy concept, mathematical idealization, and the spiritual imagination. Hinton translated the immaterial propagation of energy within the luminiferous ether as the projection of spiritual agency within hyperspace.

Hinton joined many of his contemporaries in reading Victorian physicist William Thomson's cautionary declaration of 1852—that "man as at present constituted" must submit to the rule of mechanical dissolution, the increasing entropy promised by the second law of thermodynamics—in the form of a suggestion that at some time in the future, humanity must reconstitute itself so as to reverse the universal loss of available energy. The quest for such a form of mental and technological evolution underlies Hinton's desire not just to imagine but to inhabit the fourth dimension. For Hinton, "the fundamental conditions of existence as we know it" are to be surpassed, at least in conception, by "supposing away certain limitations," in particular, those that delimit human perception to three dimensions. Hinton argued on good Maxwellian authority that physics advances by active projection beyond perceived constraints, by hypothesizing and then experimentally testing, sometimes confirming, new models of reality. The moral imperative of Hinton's hyperspace philosophy is to apply to *ourselves* this same positive, scientific technique of progressive reasoning. Augmenting the forms of spatial cognition can reconstitute humanity at a higher level of being:

> For a long time four gases were considered incapable of being reduced to the liquid state. It is but lately that a physicist has succeeded in showing that there is no such arbitrary distinction among gases. Recently again the question has been raised, "Is there not a fourth state of matter?" Solid, liquid, and gaseous states are known. Mr. Crookes attempts to demonstrate the existence of a state differing from all of these. It is the object of these pages to show that, by supposing away certain limitations of the fundamental condi-

> tions of existence as we know it, a state of being can be conceived with powers far transcending our own.[48]

Hinton returns repeatedly to the idea of the fourth dimension as a plane from which the physical and moral interiors of all three-dimensional objects are open to view. Confronted with fictive figures confined to linear, planar, or spherical surfaces, as reader one peers down from above into quaint lower worlds, to whose inhabitants one is a higher being. But in relation to a hyperspace visitor—or as A. Square was to A. Sphere in Abbott's *Flatland*—one becomes the lower being whose existence is open at all points to a panoptic higher-dimensional gaze.

> A being in three dimensions, looking down on a square, sees each part of it extended before him, and can touch each part without having to pass through the surrounding parts, for he can go from above, while the surrounding parts surround the part he touches only in one plane. So a being in four dimensions could look at and touch every point of a solid figure. No one part would hide another, for he would look at each part from a direction which is perfectly different from any in which it is possible to pass from one part of the body to another.[49]

In a later paper of *Scientific Romances,* "Many Dimensions," Hinton aimed this idea directly at "the moral life": "And so we lie palpable, open. There is no such thing as secrecy."[50] The scopic regime of Hinton's hyperspace is a paranoid structure playing between the visibility of interior spaces and the disappearance of material bodies. And yet, "a being in four dimensions would have to us exactly the appearance of a being in space,"[51] since we would see only that infinitely thin section that fell into our three-dimensional perception. One would know a hyperspace being, however, by its ability to come and go at will: "a four-dimensional being would suddenly appear as a complete and finite body, and as suddenly disappear, leaving no trace of himself, in space."[52] The allegorical subgenre of hyperspace analogism constructs unstable or evershifting spaces where the given dimensionality (or allegorical significance) of any particular "being" is in flux and thus always in question. If the phenomenal appearance of a thing is necessarily a function of the dimensionality of the mental space occupied by its observer, then anything can be transformed, simply by altering the cognitive capacities of that observer. For as Hinton now reminds us, *we* are to be the gods of this conception. Through a practical discipline of higher cognition we can deify ourselves: "Why, then, should not the four-dimensional beings be ourselves, and our successive states the passing of them through the three-dimensional space to which our consciousness is confined?"[53]

In this period of electromagnetic wonders—telegraphs and telephones—multiple technologies were coming into existence whose emerging networks offered analogous extensions of the senses. Hinton connected his scientific to his ethical theories of the spatial fourth dimension through the intermediation of a media device. The new media technologies—their promise of spiritual empowerment and their threat of technological determinism—elicited perennial philosophical issues that he resolved by contributing a notion of human volition as an inscribing of furrows on the hypersurface of the ether. Addressing the question of free will in a physical universe of ethereal materialism, he hung this ethical issue on the imagery of an exciting new technology of audio wave functions—the phonograph.

In 1880, a few years before Hinton's particular treatment of this uncanny media storage-and-delivery device, French philosopher Jean-Marie Guyau published the article "Memory and Phonograph." In this sample of phonographic technoscientism, the function of the phonograph becomes a material allegory of mental faculties—the inscribing and retrieval of memories:

> The most refined instrument (both receiver and motor in one) with which the human brain may be compared is perhaps Edison's recently invented phonograph. . . . Upon speaking into a phonograph, the vibrations of one's voice are transferred to a point that engraves lines onto a metal plate that correspond to the uttered sounds—uneven furrows, more or less deep, depending on the nature of the sounds. It is quite probable that in analogous ways, invisible lines are incessantly carved into the brain cells, which provide a channel for nerve streams.[54]

Hinton's phonographic scientism, in contrast, is cosmic rather than neurological. In classical thermodynamic scientism the steam engine is analogized to the universe; here it is the phonograph that becomes a world-machine. Hinton adapts the phonograph to the ether of space by analogizing the rotation of the audio cylinder to the revolution of the earth in its orbit, which orb then becomes both the needle plying the ether grooves and the informatic datum manifested by the vibratory registration of those indentations on the membrane of matter. With the phonograph as the mediating metaphor—the media as allegorical medium—Hinton's intelligent ether parallels Guyau's cerebral nerve-channels. Through the phonograph the cerebral microcosm communicates with the spatial macrocosm:

> For suppose the æther, instead of being perfectly smooth, to be corrugated, and to have all manner of definite marks and furrows. Then the earth, coming in its course round the sun on this corrugated surface would behave exactly like the phonograph behaves. In the case of the phonograph the indented metal sheet is moved past the metal point attached to the mem-

> brane. In the case of the earth it is the indented æther which remains still while the material earth slips along it. Corresponding to each of the marks in the æther there would be a movement of matter, and the consistency and laws of the movements of matter would depend on the predetermined disposition of the furrows and indentations of the solid surface along which it slips.[55]

Hinton's groovy phonographic ether encodes a body of data that the material universe instantiates by "playing back" as it moves over the ether-grooves. So far, however, this theory appears to entail not free will but predetermination, as with a gramophone cylinder that can only deliver the same song over and over. However, Hinton continues, "we may suppose that the æther itself is capable of movement and alteration; that it moulds itself into new furrows and marks. . . . It may be supposed that in an action of our wills we . . . may be altering these corrugations of the æther."[56] Later in this chapter we will argue that this speculation anticipates an offhand remark of D. H. Lawrence's forty years later, that "one's got to put a new ripple in the ether." Hinton's speculation also rounds out the original configuration of Edison's phonograph as a reversible device that "combine[s] the two actions indispensable to any universal machine, discrete or not: writing and reading, storing and scanning, recording and replaying. In principle . . . it is one and the same stylus that engraves and later traces the phonographic groove."[57] For Hinton, then, the human will is conceived phonographically as a potential author of as well as audience for ethereal communications, the active voice driving the stylus into the tinfoil as well as the passive ear at the loudspeaker of the Real.

Of course, this ethereal tableau of cosmic volition could not be farther from Kittler's trenchant analyses of media determinism. Because "the bodily real . . . of necessity escapes all symbolic grids," Kittler writes, "media always already provide the appearances of specters."[58] That is, any mediated reproduction of a material phenomenon determines the separate and incommensurate existence of a phantom proxy, a ghostly data-body. The fourth dimension that descends from Hinton to Lawrence, from late-classical hyperspace to early modernism generally, appears to be an Imaginary space that recuperates the noise of the Real. Hinton's fourth dimension of space is to some degree a spectral reinscription of the electromagnetic ether, which is itself, in Hinton's exposition, an optical reinscription of the acoustical mechanics of the phonograph.

Once the phonograph interposes itself for real into our world as a media device, it drives us to envision the world in its entirety as a phonograph cylinder, a storage-and-transmission device, a consortium of natural and cultural objects engraved with informatic grooves. Thus, on the one hand, as we have seen, in the 1880s Hinton seized the ether of space as a phonographic surface. And on the other hand, in a gesture corresponding to Guyau's article "Memory and Phonograph," in 1919 the poet Rainer Maria Rilke wrote of constructing a crude phonograph in a schoolboy science class, and then, as a grown man, of catching a passing glimpse by candlelight of a human skull: "the coronal suture had become strikingly visible, and I knew at once what it reminded me of: one of those unforgotten grooves, which had been scratched in a little wax cylinder by the point of a bristle!"[59] Kittler comments on this passage: "What the coronal suture yields upon replay is a primal sound without a name, a music without notation . . . . A writer thus celebrates the very opposite of his own medium—the white noise no writing can store. . . . Bodies themselves generate noise. And the impossible real transpires" (44–46).

Kittler aligns the three realms of Lacanian discourse—the Real, the Imaginary, and the Symbolic—with the media triumvirate of gramophone, film, and typewriter. In this scheme the phonograph not only corresponds to the real, it also constitutes it as a cognizable and manipulable, if unknowable and noumenal, substance. "Thanks to the phonograph," Kittler writes, "science is for the first time in possession of a machine that records noises regardless of so-called meaning."[60] Kittler's gramophone busts open the realm of the real, the nonsymbolic and nonimaginary residues of mediation. Mechanical inscription—storage and transmission—will now reorganize human consciousness—memory and discourse. The phonographic imaginary, I would add, already adumbrates the mechanomorphic cyborg. We moderns imagine our own unimaginable implications in the noise and mire of material nature in the form of technomorphoses of our mediated bodies. In the cultural work of the psychic ether and the spatial fourth dimension, allegories of media are being joined to allegories of energy. In the transformation of the ether medium into the spatial and temporal fourth dimensions, mediated reality is metamorphosed into art forms—imaginary realms and symbolic structures.

When Hinton's ether discourse modulates from physical theory to the parapsychology of a self-determining personality, the idealism underlying his hyperspace investigations breaks entirely loose.[61] He applied the ether to a theory of the soul, which becomes in this idiom the "aethereal body"—a spectral data-body inscribed within the ether as informational matrix. Hinton developed this idea in a manner consistent with his approach to the fourth dimension of space. Like the hyperbeings that emerge in the interstices of his fictive dimensional narrations in *Scientific Romances,* here he reified the ethereal body as a higher form of being, an eternal ether-being absolved from its fortuitous connection to the mortal material body that it encodes. Therefore the proper human attitude toward it must be one of solicitude for the higher unity to which we owe our span of 3-D bodily existence. The transcendental ether body can determine its own grooves:

> In the supposition of an aethereal body, it is not necessary to keep to the idea of the rigidity and permanence of the grooves defining the motion of the matter which, passing along, exhibits the material body. The aethereal body may have a life of its own, relations with other aethereal bodies, and a life as full of vicissitudes as that of the material body, which in its total orbit expresses in the movements of matter one phase in the life of the aethereal body.[62]

Hinton then backed away from this cosmic prospect to remind the reader that "these speculations . . . are only introduced here in order to show how the conception of higher space lends itself to the representation of certain indefinite apprehensions."[63] The overriding purpose in elaborating his ether models has been to inspire the reader to undertake the practical and arduous discipline needed to train one's higher space-intuition. Hinton had to assert so strenuously the freedom and self-commanding effectivity of the will in order to call the fourth dimension forth, to shape and expand the individual mind to its higher potential and unity. Ultimately, then, for Hinton the fourth dimension of space was not really an undiscerned region of the physical universe, as the ether was taken to be an undiscerned substance. It was not a discovery waiting to be confirmed so much as a creative ideal—a cultural prophecy and moral goal to which the mind ought to be conformed.

Hinton used dimensional analogies to establish the fourth dimension's plausibility, but he also confessed that full or literal access to four-dimensional cognition could result only from an arduous intellectual discipline, for which he had developed a stunningly intricate set of mental exercises. The readable portions of his work are essentially prolegomena for and inducements to the further, impenetrable system of hyperspace-instruction. Hinton's weird manual in *A New Era of Thought* for training higher space intuition with vast sets of multicolored cubes offers practice in the four-dimensional rotations of a tesseract, or hypercube.[64] There is an arduous playfulness here: one might also speculate that in these documents an obsessive-compulsive manipulation of childhood objects defends against a paranoid cognition of being psychically manipulated from another source of control. But Hinton's quest for a discipline to induce a cognition of the fourth dimension was also a specific and significant response to the evolutionistic vogue for superhuman types—from Nietzsche's *Übermensch* to Dora Marsden's Freewoman—at large in the later nineteenth and early twentieth centuries. Hinton's fourth dimension of space offered a progressive vista of vital and material transformation further developed early in the twentieth century, when the creative possibilities of higher dimensionality, "this new kind of depth," were unfolded in the experimentalist aesthetics of modernist art and literature.[65]

In sum, Hinton's writings offer a concrete resource for a cultural poetics of scientific, mathematical, and philosophical ideas. His particular textual constructions of the fourth dimension divulge a number of discourses that contributed to the vogue for the fourth dimension in the intellectual and aesthetic circles of the earlier twentieth century. For instance, as we will note in the next section, Hinton's notions about laying down new grooves in the ether seem to have enjoyed a certain cultural currency, at least among the more cosmical wing of Anglo-American literary modernism. At times Hinton's writings are hinged directly to authoritative sources in Victorian physics and mathematics. At other times they operate a form of post-Kantian Platonism to produce ethereal imaginations similar to the cosmologies of concurrent spiritualistic thinkers eager to claim the scientific and technological developments of their day in support of esoteric beliefs. However, Hinton's desire for scientific credibility precluded the full-blown occultism of Madame Blavatsky, Edward Carpenter, or the symbolist poets. While beyond the scientific pale, his speculations produced a minigenre of scientistic space-fictions with some significant relations to the nascent literature of science fiction. In the connections of the luminiferous ether to Hinton's fourth dimension of space, a significant Victorian allegory of energy emerged through a "crack in the walls" of Euclidean geometry and began to transform the discourse and practices of modernist culture.

Lawrence's Ether

In the decades leading up to Anglo-American and European modernism, the physical ether gave the transcendentally inclined a scientistic channel for metaphysical communications. "Looking back upon it now," Mabel Dodge Luhan wrote in her memoir *Movers and Shakers,* published in 1936, "it seems as though everywhere, in that year of 1913, barriers went down and people reached each other who had never been in touch before; there were all sorts of new ways to communicate, as well as new communications."[66]

Movers and Shakers evokes the transcendental element of the early modernism abroad in the cosmopolitan art and intellectual scenes of the time: "Men began to talk and write about the fourth dimension, interchangeability of the senses, telepathy, and many other occult phenomena without their former scoffing bashfulness, only they did it with what they were pleased to call a scientific spirit. The essence of it all was communication."[67]

Mabel's unabashed and unorthodox spirituality was not a particularly marginal point of view. Rather, the doctrinal foundations of her spiritual quest embrace many of the aesthetic philosophies that informed the early modernist movements of 1910s. Her remark that the "new spirit" was advanced with "a scientific spirit" reminds us, as Linda Dalrymple Henderson has demonstrated at length, that at the turn of the twentieth century, the new spectrum of radiant energies and their technological applications—telephony, wireless transmissions, x rays, and so on—were immaterial phenomena broadly susceptible of metaphysical interpretations. By the end of the nineteenth century, the solid cosmos in its entirety was being redescribed precisely as a *wave* phenomenon.[68] And for many besides Mabel, the vibrations, pulses, and undulations of aural, optical, and electromagnetic waves were taken to demonstrate at the physical level and so ratify the more rarefied but equally wavy mechanisms of psychic phenomena.

A century ago the development of the telegraph, photograph, telephone, and phonograph were as profoundly suggestive as contemporary digital technologies are for us. For instance, in Blavatsky's *Isis Unveiled,* the exciting news of "the possibility of annihilating distance between human voices—by means of the *telephone,*" is received as a technological realization of spiritual capacities.[69] Thirty years later, in *The Art of Creation: Essays on the Self and Its Powers,* Edward Carpenter articulated a form of visionary scientism that epitomizes the sort of common cultural source both Mabel Dodge Luhan and D. H. Lawrence drew on for their own notions of communication and cosmic relation. With the following daemonic vision of the new telecommunicating cosmos, Carpenter depicts the modern mythopoetic overlapping of telepathy and technoscience:

> Nature is a great vehicle, an innumerable network and channel of intelligence and emotion. . . . Countless hosts of living beings, of every grade of organisation and consciousness, are giving utterance to themselves. . . . With incredible speed the messages of these intelligences flash through space; "the Morning Stars sing together"; the messages of light and sound and electricity and attraction penetrate everywhere; and as modern science shows us that the air, the sea, and the solid frame of the earth itself may be the vehicle of waves which without wire or definite channel may yet convey our thoughts safely to

> one another through intervening leagues of distance, so surely we must believe that the countless vibrations ever going on around, and ever radiating from and impinging on every known object, are messengers too of endless meaning and feeling.[70]

The modernist vision of telepathic communication or extrasensory perception is a cultural successor and parallel to the Victorian spiritualist movement. It reflects in a mythopoetic mirror the technoscientific arrival of telegraphy, telephony, and radio. In her evocation of the new spiritual communications that mix science and mysticism, Luhan got right what such modern mythopoeses do get right—the vital significance of the elaboration and exchange of *information* in an evolving universe of matter and energy.[71]

But because the ether was also, and originally, a scientific concept adapted from classical Greek ideas to the post-Newtonian cosmos, the modern mystic concept of the ether rests, paradoxically, on a mechanistic science whose premises demanded a material substance or physical medium by which to explain the operations of gravity and radiation.[72] One faithful and fateful follower of Newtonian mechanics was Franz Anton Mesmer, whose invention of animal magnetism in 1776 had such remarkable repercussions, such as psychoanalysis and many other systems and fantasias of the unconscious. Mesmerism postulated a universal substance that Mesmer himself distinguished from the ether per se, but both conceptions are essentially identical.[73] Mesmer's universal fluid, however, crossed over from physics to physiology as "a principle acting on the nerves."[74] And from here the way was clear to give a plausible scientific explanation to "the communication of will." In verbal and visual communication, speech and gestures are "transmitted to the air or to the ether—as intermediate mediums—in order to then be received and sensed by the external sense organs"; at the same time, our thoughts and desires are transmitted to and by the universal fluid that binds us to everyone and everything: "We can thus comprehend how the wills of two persons can communicate with each other through their internal sense organs; consequently, we can understand how there can exist a reciprocity, an accord, a sort of 'covenant' between two wills, which we call 'being in rapport.'"[75]

The common element connecting the Mesmeric fluid and the modernist ether is a subtle materiality sufficient to penetrate to the pith of an organism and thus "beat" on the very pulses of the "living plasm," as Lawrence would say. Mesmer's universal fluid in "continuity" with the "nervous substance" distinctly anticipates the kind of telepathic vitalism we will find in Lawrence's particular systems of ethereal communication.[76] In *Lorenzo in*

Taos, for instance, Mabel was forthright about her exertions of psychic willpower to make the ethereal connection she hoped would draw Lawrence to her residence in New Mexico: "Before I went to sleep at night, I drew myself all in to the core of my being where there is a live, plangent force lying passive—waiting for direction. Becoming entirely that, moving with it, speaking with it, I leaped through space, joining myself to the central core of Lawrence. . . . I became that action that brought him across the sea. 'Come, Lawrence! Come to Taos!' became, in me, Lawrence in Taos."[77] And for whatever reason (she had also sent him encouraging letters by international post), Lawrence arrived. Not surprisingly, he soon became critical of Mabel's mystic will. By casting Mabel in particular as a cautionary figure of "willfulness," Lawrence's later diagnoses of modernist malaise centered not precisely on ordinary neurosis but rather on the *spiritual* presumption involved in the invocation of occult powers for the gratification of egoistic motives.[78] The direction of volition is the flash point in Lawrence's quarrels with "egoists" as well as mystical idealists. In the fiery strife of wills that developed between Luhan and Lawrence the main item of contention was the *proper* cosmic application and issue of individual volition.

Lawrence articulates his critique of Mabel's mystic will with particular force in two remarkable letters she reprinted in *Lorenzo in Taos.* In the first, he wrote:

> Surely it is even a greater mystery and preoccupation even than willing, to let the invisible life steal into you and slowly possess you. Not people, or things, or action, or even consciousness: but the slow invasion of you by the vast invisible god that lives in the ether. . . . Instead of projecting your will into the ether of the invisible God, let the invisible God interpenetrate into you.[79]

Here is an idiom of spiritual communication similar to *Movers and Shakers* and Edward Carpenter's *Art of the Self.* Lawrence's theologized ether medium enables human and divine intercommunication, in the context of the different attitudes the human will can take toward its acknowledged capacities to induce psychical effects by transmitting or receiving the frequencies of ethereal waves. In a second letter three months later, Lawrence bid Luhan to alter the outcomes of her spiritual force, not by changing the objects of her will, but by desisting altogether from the forcing of volition either to dominate or to submit. Sounding uncannily like Charles Howard Hinton, Lawrence implored Luhan to "put a new ripple in the ether." But Lawrence's animus is directed precisely at the same sort of transcendental volition that overcharged Hinton's cognitive quest for the fourth dimension of space: "Not any forced will of your own, nor any forced submission, but a certain *real* trust, and the courage *not to care,* and the power to laugh a bit. Do this and we'll have a good time among ourselves. One's got to put a new ripple in the ether" (129). Lawrence's injunctions amount not to an abandoning of mystic presumptions and presuppositions, but to their redirecting from transmission to reception. As he had made explicit in his first letter: "it's only a change of direction" (118).

The idiom of a spiritual ether is no momentary usage or rhetorical ploy on Lawrence's part. On the contrary, this line of electrovitalism represents a fundamental cosmic fixture informing his prophetic pronouncements about the ways and means of what he called "vital relation." In *Women in Love,* completed in manuscript in 1916, the capacity of the ether for transmitting ineffable and potentially deadly messages is darkly intimated when Gerald has a visionary response to "long red rent" that Bismarck the rabbit inflicts on Gudrun's forearm: "The long, shallow red rip seemed torn across his own brain, tearing the surface of his ultimate consciousness, letting through the forever unconscious, unthinkable red ether of the beyond."[80] The first version of his essay "The Spirit of Place," published in November 1918, figures the ether more positively and distinctly, when Lawrence explains Aztec prophecies concerning "the coming of the white stranger" as the reception of ethereal transmissions. The Aztecs' "consciousness was fluid, not mechanically fixed, and the rarest impressions upon the physical soul, from the invisible ether, could pass on occasionally into uninterrupted consciousness."[81]

In his essay "The Two Principles," published in June 1919, Lawrence rejected the mechanistic cosmology of "the modern, scientific attitude," because "Science supposes that once the first forces were in existence, and the first motions set up, the universe produced itself automatically, throwing off life as a by-product."[82] This is a reasonable description of the deterministic cosmos of classical or Newtonian science against which vitalists had been rebelling for several centuries.[83] Anticipating his cosmological claims in *Fantasia of the Unconscious,* Lawrence posited Life's incorporation in a kind of divine nebular fluid, in advance of the creation of physical matter with determinable motions: "When God made Heaven and Earth," Lawrence intoned, "we must conceive [of him as] some homogenous rare *living* plasm, a *living* self-conscious ether, which filled the universe."[84] In this visionary construction of cosmic origination, the Lawrencean ether is both theologized and vitalized. God and Life become interchangeable within a polarized electromagnetic medium.

The Lawrencean ether also animates poems collected in *Birds, Beasts and*

Flowers but written before the advent of Mabel's direct influences. For instance, "Bare Almond Trees" and "He-Goat," composed in Sicily in 1920–21, lock together to evoke the electromagnetic ether in its cosmic function of informatic propagation. "What are you doing in the December rain?" the poet asks the almond tree:

> Have you a strange electric sensitiveness in your steel tips?
> Do you feel the air for electric influences
> Like some strange magnetic apparatus?
> Do you take in messages, in some strange code,
> From Heaven's wolfish, wandering electricity. . . ?[85]

Lawrence figures the almond tree as a radio antenna, but conducts the mechanical metaphor toward the vital vibrations that establish the almond tree's relatedness to a living cosmos. In "He-Goat" the creature's tail also becomes an antenna, and Lawrence names the ether explicitly as the medium of the creature's sexual energy: "from the stem of his bristling, lightning-conductor tail / In a rush up the shrieking duct of his vertebral way / Runs a rage drawn in from the ether divinely through him / Towards a shock and a crash and a smiting of horns ahead" (157). *Birds, Beasts and Flowers* details a living creation that evolves from and is interpenetrated by a divine ether bearing prowling waves of "wolfish, wandering electricity."

In a fine passage from the essay "Indians and Entertainment" of spring 1924, written during his second sojourn in New Mexico and collected in *Mornings in Mexico,* Lawrence wove into the aboriginal tableau of a New Mexican Indian Corn Dance his vision of the vital waves that connect a communal human volition in pristine purpose with the ever unfolding re-creation of the living world. In these formulations Lawrence reinscribes the electromagnetic ether and its mechanical offspring into the register of vital energies. This scientistic ether propagates pulsating waves of electrovitalism:

> The drum is a heart beating with insistent thuds. And the spirits of the men go out on the ether, vibrating in waves from the hot, dark, intentional blood, seeking the creative presence that hovers forever in the ether, seeking the identification, following on down the mysterious rhythms of the creative pulse, on and on into the germinating quick of the maize that lies under the ground, there, with the throbbing, pulsing, clapping rhythm that comes from the dark, creative blood in man, to stimulate the tremulous, pulsating protoplasm in the seed-germ, till it throws forth its rhythms of creative energy into rising blades of leaf and stem.[86]

What do Lawrence's remarks have to do with the advent of modern technological media? The answer lies in that "new ripple in the ether." This means perhaps an act of cosmic volition, the bringing of something new into the world. But his choice of rhetorical vehicle alludes to the propagation of waves within a communications medium. Lawrence's proximity to Mabel Dodge Luhan, it would seem, only drew out of him in more explicit fashion an ether idiom that was already deeply embedded in his residual vision of the cosmos. The intensely "electromagnetic" tenor of Lawrence's writings has been noticed before, but not to my knowledge fully coordinated with the ether of space nor with the organic component of his particular brand of ethereal electrovitalism. In the nineteenth century, Mesmer's electromagnetic fluid converged with the wave theory of light and the luminiferous ether posited by Victorian physics as the single material carrier of both optical and electromagnetic undulations. This vitalistic allegory of vibrating space grounds the total circulation of electromagnetic metaphors in Lawrence's scientistic visions of cosmic communication.

For the popularizers of Victorian science, light and electromagnetic radiation were the waves that rippled within a sea of ether filling up all space. When radio was invented, the "air-waves" carrying its disembodied voices from station to station were construed as vibrations of the same ether. Conveniently for the elaboration of visionary scientisms, the ether of space remained virtually occult—no definitive evidence of its real existence would ever be found. In the decades leading up to the generation of Mabel Dodge Luhan and D. H. Lawrence, the ether provided a wide cultural medium of cosmic reconciliation: it gave theological scientists a physical bridge to heaven, while it gave occultists seeking social validation a niche within the discourse of authorized science. So the ether passed back and forth from a physical plenum to a divine omnipresence. Transcendental modernism carried on this powerful cultural amalgam of natural science and metaphysical doctrine.

Chapter 8
Literary Relativity

Concepts of Physical Relativity

Although his first scientific paper on the special theory of relativity had been published in 1905, Einstein's broad international renown emerged only after World War I. His fame was sealed by the solar eclipse of 1919, when British physicist Sir Arthur Eddington provided empirical confirmation of the general theory of relativity by documenting the bending of starlight passing through the gravitational field of the sun.

Einstein's 1916 monograph *Relativity: The Special and General Theory, a Popular Exposition,* was issued in English translation in 1920. Early in this exposition, in order to set the stage for his special and general theories of relativity, Einstein introduced the *principle* of relativity, a physical theory first defined by Galileo and a standard part of the Newtonian or classical picture. The principle of relativity states that when two reference frames or coordinate systems *K* and *K´* are in "a condition of uniform motion of translation," that is, moving with respect to each other at a steady rate without acceleration or rotation, "then natural phenomena run their course with respect to *K´* according to exactly the same general laws as with respect to *K*."[1] So a limited form of relativity already existed in classical mechanics: for certain well-defined relations of motion, there is no alteration of natural laws as one passes between one frame of reference and another. Einstein's special and general theories of relativity dramatically extend the range of lawful equivalences among physical systems, with the effect of compressing the possibilities of cosmic phenomena into newly manageable mathematical formulas. Einstein's conclusion: "The great power possessed by the general principle of relativity lies in the comprehensive limitation which is imposed on the laws of nature" (111).

The principle of relativity exhibited by so-called Galilean coordinate systems is clarified by Einstein's famous thought-experiment of a railroad where two separate observers occupy different reference-frames. One observer is in motion on a train traveling at a uniform velocity along a rectilinear track, and the other is standing on a footpath beside the railway embankment to which the track is fastened. Einstein now imagines that he is the observer in the moving car: "I stand at the window of a railway carriage which is travelling uniformly, and drop a stone on the embankment, without throwing it. Then, disregarding the influence of the air resistance, I see the stone descend in a straight line. A pedestrian who observes the misdeed from the footpath notices that the stone falls to earth in a parabolic curve. I now ask: Do the 'positions' traversed by the stone lie 'in reality' on a straight line or on a parabola?"[2] The answer depends of course on one's choice of perspective, that is, on where one fixes one's reference frame, the train traveler's moving carriage or the pedestrian's footpath. Although these different observers witness a different form of reality, either is equally real, and the laws of nature do not change in order to produce these different perceptions.

For example, at one time the earth itself was conceived to be resting motionless at the center of creation, and relative to this fixed reference frame, the sun and the heavens were perceived as describing huge motions, diurnal revolutions and annual variations. While we now understand conceptually that in fact, the earth rotates on its axis as well as revolves around the sun, the principle of relativity reminds us why we still tend to experience the sun as in motion and ourselves as stationary. When observations are made from a moving body, those motions will be projected onto objects outside that frame of observation. From a moving car, for instance, the world appears to be rushing by, while the coffee I hold in my hand appears to be perfectly still. Once we acknowledge the actual trajectory of our own standpoint, as Copernicus did with regard to the earth, we can begin to compensate for the relative limitations of our particular perspectives, and in doing so we increasingly discover the unity rather than divergence of natural laws.

Let us transpose these considerations to the universe at large. We are being wheeled around by the earth's diurnal rotation and annual revolution, and further, the sun itself is being carried around the axis of the Milky Way once every 400 million years or so, such that the earth's revolution describes not an ellipse but a complex ellipsoid spiral. Add to that the velocity of the Milky Way as it participates in the expansion of the universe begun 10 or 20 billion years ago by the Big Bang and detectable by the red shift in the spectral lines of distant and receding galaxies. What is the virtual velocity of the earth's movement through space?

The theory of relativity affirms that we are streaming through space at an incalculable velocity *and* we are stock still, and in either case, the same laws of nature hold. Our question makes no physical sense unless a fixed reference point is posited upon which to initiate a measurement. In our universe, the fixing of that point can only be arbitrary. One frame is as good as the next. The question of absolute velocities simply has no determinable answer, because the cosmos possesses no absolute standpoint at perfect and utter rest relative to everything else.

However, the modern relativistic perspective was announced to a world long invested in physical and metaphysical convictions about the ultimate reality of some form of absolute reference. In classical physics this "eternal axis" was identified with the hypothetical ether of space, which entity was presumed to be at rest in relation to the planets and stars. "The notion of an ether which pervaded the entire universe, being everywhere at rest in some particular frame of reference . . . lent support to the idea of an *absolute* frame of reference, with respect to which *the absolute position and velocity of all bodies could be specified.*"[3] Einstein's popular exposition deals curtly with this aspect of the ether-concept: the special theory of relativity asserts that "there is no such thing as a 'specially favoured' (unique) co-ordinate system to occasion the introduction of the aether-idea."[4]

We continue to cross over from the principle of relativity to the modern theory of relativity once we factor the relativity of *time* into the relativity of motions in space—that is, once we begin to think in terms of a four-dimensional continuum composed of three dimensions of space and one of time. In the relativistic cosmos, time is coupled to space most profoundly in *c*, the velocity of light—186,000 miles of space per second of time—first determined in the mid–nineteenth century. With regard to the physics of light, Einstein's new theories had to deal with a number of observational problems for which classical mechanics had no answer. It was previously unaccountable that the speed of light is both finite and constant, whereas in our more familiar Newtonian world velocities are variable and compoundable. From a high-classical perspective, the constancy of the speed of light is counterintuitive. Although it defies the additive property of classical velocities, as in the compounding of cosmic motions we just discussed, "The velocity of propagation of light cannot depend on the velocity of motion of the body emitting the light."[5]

The finitude of *c* also means that the spatiotemporal relations of different coordinate systems are not absolutely simultaneous but relative to their respective velocities and to the distances between them. Einstein comments that "in the theory of relativity the velocity *c* plays the part of a limiting velocity, which can neither be reached nor exceeded by any real body."[6] The theory of relativity predicts that if any portion of matter were to achieve the velocity of light, its mass would become infinite. Although this never happens, it means that mass is relative to velocity, which is relative to the time and energy necessary to accelerate the body in the first place. In addition to the relativity of space and time, then, and in a fashion unpredicted by classical conservation laws, mass and energy are also relative to each other, as captured in Einstein's most famous equation, $E = mc^2$. The finite constancy of light provides the theory of relativity with mathematical tools and physical principles for reconceiving the functional equivalence of spatiotemporal reference frames with regard to all velocities and all possible rotations. Space and time are thus woven into a single cosmological fabric with gravity and electromagnetism.

Edward Carpenter's "Modern Science: A Criticism"

The lapse of the ether out of authorized science was received in different ways in various discursive quarters. The broad reluctance to dispense with it and thus with other reassuring continuities of the classical world picture was expressed most acutely in the later career of Sir Oliver Lodge.[7] Others' responses were more mixed. For the lay observer of science Edward Carpenter, for instance, the physical problematics associated with the ether had been grounds for an equally problematic philosophical critique of scientific dogmatism. When the spatial conundrums of the ether were finally unraveled by relativity theory, the results were not entirely what Carpenter would have predicted.

Carpenter's contemporary Charles Howard Hinton explicitly marked his gestures toward a possible physics of the fourth dimension with the heuristic moves of Clerk Maxwell's development of electromagnetic field theory. Whereas Hinton's discourse traded on the cultural authority of Maxwell's science, in his critique of that same science Carpenter used Maxwell, quite unfairly, as a metonymy for a blinkered scientific materialism. The case of Carpenter is especially interesting because he was in other respects an assiduous amateur student and humanist expositor of late-Victorian science. But his bias was clearly toward the life sciences in their neo-Lamarckian and early eugenic formations, and insofar as they could be taken to lend weight to his mystic evolutionism. In contrast, discourses of energy often appeared to offer little but dissipation and heat death. So Carpenter heaped upon physics and physicists a generalized opprobrium aimed at "mechanistic" methodologies and the hyperbolic truth claims of "modern science" in general.

Carpenter's *Civilization: Its Cause and Cure* went through fourteen editions between 1889 and 1921, the period of fundamental transition from the paradigm of classical dynamics and ether theories to the paradigm of Einsteinian space-time and the regimes of special and general relativity.[8] Thus one essay in particular from that collection, "Modern Science: A Criticism," came to straddle two distinct eras of physics. In the thirteenth edition, published in 1914, its text is unchanged from the 1889 version. Only with the final, "newly-enlarged and complete edition" of 1921 did Carpenter address the sea change in the meaning of "modern" science, and even then, only in certain limited respects. In May 1920, Carpenter wrote his old friend Havelock Ellis, who had reviewed the book for a last reissue, to remark that he had found only a few things in the previous edition that demanded attention.[9] As we will note below, one of these alterations concerned the advent of Einstein's physics.

In 1921 as in 1889, "Modern Science: A Criticism" begins with a satirical personification of Science as a questing spirit now perplexed by complexity: "After a glorious burst of perhaps fifty years, amid great acclamations and good hopes that the crafty old universe was going to be caught in her careful net, Science, it must be confessed, now finds herself in almost every direction in the most hopeless quandaries."[10] Carpenter drew up a list of research corridors that in his estimation had failed in their positive goals. Prominent on that list are two glaring deficiencies in the area of physics: "Indeed the state of the great theories of Science in the present day—the confusion in which the Atomic theory of physics finds itself; . . . the appalling and indeed insurmountable difficulties which attach to the Undulatory theory of Light . . . —all these things do not seem to point to very near limits of rigorous exactness!" (59–60). In 1889 Carpenter had some justification to underscore the unsatisfactorily hypothetical status of atomic theory and of the doctrine of the luminiferous ether: "in the case of the undulatory Theory of light or the Atomic theory of physics, the concepts thus erected into actualities are composed of purely imaginary attributes—of which no one has had any experience—an undulatory ether in one case, a hard and perfectly elastic atom in the other" (70). Thirty years later, the rise of relativity and quantum mechanics and the collapse of classical absolutisms had swept away Carpenter's justifications as well as his own, residually mechanistic complaints about the heuristic modeling of "imaginary attributes." Yet these passages remained unchanged in the 1921 "newly enlarged and complete" edition.

The breakup of the classical physical paradigm led to a progressive relinquishing of "pictures" such as Maxwell's electromagnetic ether model and to their replacement by mathematical constructions. As Sir James Jeans wrote in 1930 in celebration of the new mathematical physics that had superseded the mechanistic cosmos of the Victorian "engineer-scientists": "The concept of energy flowing about through space is useful as a picture, but leads to absurdities and contradictions if we treat it as a reality. . . . The mathematician brings the whole problem back to reality by treating this flow of energy as a mere mathematical abstraction."[11] In other words, the problem with the reification of the ether was not methodological, as Carpenter claimed, but paradigmatic: it was rooted in the mechanistic demand for some form of material continuity by which to force an energy flow across space: "Our difficulties have all arisen from our initial assumption that everything in nature, and waves of light in particular, admitted of mechanical explanation: we tried in brief to treat the universe as a huge machine" (96).

At this point in Carpenter's 1921 text of "Modern Science: A Criticism," however, a short sequence of enlargements commences. The first is a footnote attached to the foregoing critique of physical concepts, including the law of gravitation: "See the report of the joint meeting of the Royal Society and the Royal Astronomical Society, November 6, 1919, when Einstein's theory was discussed."[12] This note marks the most inescapable development in the physical department of "modern science," one that rouses the author to a modest effort of updating. Carpenter's papers contain the page proof of a second, more extensive footnote on the theory of relativity, which he then revised and extended further into the version that appears in the published volume of the fourteenth edition:

> It is obvious that the Einstein theory, in which Time enters as a kind of fourth dimension in relation to Space, removes us at once out of the whole field of ordinary scientific reasoning and lands us, so to speak, in a new world. The nature of Space (or of the universal medium, whatever it is) in any region—its possible fundamental accelerations there, its "curvature" or non-Euclidean character, and so forth—is supposed, according to this theory, to vary with the amount of matter in, or density of, that region; and the movements of bodies are consequently supposed to take on the characters (accelerations, etc.,) which we ascribe to the action of Gravitation. Gravitation in fact in any region is the manifestation in Time of the attributes of the universal Medium in that region—which latter again is dependent on the degree of Matter present. Thus, Matter, Time, and Space are *one phenomenon.*
>
> The whole Einstein theory, in fact, is a device to present these three Protean and variable elements of all material existence (Matter, Time, and Space) as so far involved and interlaced in each other that they form always an absolute and complete unity. As such the theory is no doubt suggestive, and along the line of future speculation: but it awaits corroboration. If corroborated it will point the way to a new conception of the Universe.[13]

Carpenter's synopsis of relativity is reasonably accurate, but also, and characteristically, he seized on the Einstein vogue for the possibility of a newly intuited "mystery science" that will work a miraculous cultural transformation.[14] His redacted version of the theory of relativity enables us to see how neatly the advent of space-time fulfilled preexisting cultural categories, directing the mystic anticipation of a fourth dimension of space toward the material and energetic language of gravitational field theory. Yet Carpenter appeared not to credit the body of experimental corroboration already achieved by 1920—in particular, Eddington's 1919 demonstration. Most tellingly, Carpenter did nothing to indicate his awareness that "the Einstein theory" had mooted precisely that aspect of classical physics under critique in the 1889 version of his essay—the problem of the luminiferous ether. By 1921 the object of his primary indictment of "modern science" has been duly removed from mainstream consideration, rendered conceptually defunct by the very developments that he had belatedly imported into footnotes to maintain a modicum of scientific currency. Although by 1920 the Victorian ether of space had departed the scientific stage, his critique of its physical difficulties remains standing alongside his new Einsteinian documentation.

Carpenter's curious polemic against the ether in the original version of "Modern Science: A Criticism" documents the doctrinal cross-purposes that could inhabit the late-classical unwinding of that concept. Whereas most of his transcendentalist colleagues had welcomed Victorian ether-science as a corroboration of metaphysical convictions, and Carpenter himself plied the rhetoric of ethereal communications, as in the passage from *The Art of Creation* cited in the previous chapter, in this text he set up his larger polemic by castigating the ether-concept as *too* metaphysical, too removed from commonsense empiricism. Thus Science's demand for popular credence in this hypothetical medium was of a piece with its cultural arrogance and unwarranted prestige:

> Again, the undulatory theory of Light. . . . There is no harm in the assumption of vibration as long as it is understood to be a mere assumption for a temporary convenience of thought. But now Science goes farther than this, and not only supposes a common attribute to all visible bodies, but credits this common attribute with a real existence independent of the visible bodies in which it was supposed to inhere—and makes this the *cause* of their visibility! Obviously now a common and universal medium is required for this common and universal assumed vibration (just as Newton required a medium for his universal "falling")—and so, hey presto! we have the Undulatory Ether. And having got it we find that to fulfill our requirements it must have a pressure of 17 million million pounds on the square inch, and yet be

> so rare and tenuous as not to hinder the lightest breath of air . . . and so on and on! In fact, we find that it is unthinkable. Against this adamantine, impalpable Ether, as against this instantaneous, untranslatable gravitation, Science bangs its devoted head in vain. Having created these absurdities by the method of "personification of abstractions" or the "reification of concepts," it seriously and in all good faith tries to understand them; having dressed up its own Mumbo Jumbo (which it once jeered at religion for doing) it piously shuts its eyes and endeavors to believe in it.[15]

A footnote to the cited phrase "reification of concepts" tips Carpenter's hand toward the commentator upon whose critical authority he has been launching these characterizations of improper scientific method, the American jurist and philosopher J. B. Stallo and his 1882 volume *The Concepts and Theories of Modern Physics.* Here Stallo launched a "common-sense" critique of the conceptual pretensions of the "atomo-mechanical theory" as a sort of metaphysical hangover, in particular, as a form of reification allegory. For Stallo (as for the philistine wing of scientific commentators to this day), any recourse to the mythopoetic was a lapse into the prescientific. The main culprits on Stallo's list are, not surprisingly, Maxwell, Stewart, and Tait: "But even the intellects of men of science are haunted by pre-scientific survivals, not the least of which is the inveterate fancy that the mystery by which a fact is surrounded may be got rid of by minimizing the fact and banishing it to the regions of the Extra-sensible. . . . All such attempts are nugatory; the intangible specter proves more troublesome in the end than the tangible presence. Faith in spooks (with due respect be it said for Maxwell's thermo-dynamical 'demons' and for the population of the 'Unseen Universe') is unwisdom in physics no less than in pneumatology."[16]

Interestingly, the excerpt Carpenter gave from Stallo also addresses the latter's treatment of "transcendental geometry" as another example of misplaced "realism" in the physical sciences:

> Even the science of mathematics—the exactest of all the sciences, whose methods are said to be as infallible as its foundations are supposed to be permanent, and which, ever since the dawn of human intelligence, has pursued the even tenor of its way amid all the vicissitudes of speculation—is not exempt from the prepossessions of ontological realism. The same hypostasis or reification of concepts, which has given rise to the atomo-mechanical theory in physics, has led to the doctrine of pan-geometry in mathematics. The hypostasis of space, by the mathematicians, is a strict analogue of the hypostasis of mass and motion by the physicists.[17]

That is, Stallo clearly cognized but did not fully appreciate either the heuristic possibilities or the cultural significations of these allegorical constructions. Stallo grasped but could not make acceptably scientific sense out of the crucial connections of thermodynamic and electromagnetic processes to the new mathematics—specifically, the crossover between the luminiferous ether and the fourth dimension of space. Being neither clairvoyant nor uncommonly prescient, he could not foresee that the emergence of "the new geometrical faith" throughout the nineteenth century would be the most profound foundation of the new physics of the twentieth century. Inviting his reader to share his own appalled intellectual sensibilities, Stallo closely sketched the outrageous non-Euclidean notions already abroad in the 1880s:

> The articles of the new geometrical faith are certainly startling. Among them are propositions such as these: that our ordinary "Euclidean" tridimensional and "homaloidal" (flat) space is but one of several possible forms of space; . . . that true and real space . . . has, or at least, for aught we know, may have, not three but four or even a greater number of dimensions; that the space in which we move is, or may be, not homaloidal or flat, but essentially non-homaloidal, curved, spherical or pseudo-spherical, so that every line, which we have hitherto regarded as straight, may upon sufficient prolongation prove to be a closed curve; . . . that not only the measure of curvature of space, as well as the number of its dimensions, may be, and probably is, different in different spatial regions, so that no valid inference can be drawn, from our experiences in the regions in which we happen to dwell, as to the curvature or the dimensions of space immeasurably distant or immeasurably small, but that in any given region both the curvature of space and the degree or number of its dimensions may be, and probably is, undergoing a gradual transformation, and so on.[18]

With a century of hindsight, it is fascinating to reflect that the mathematical and physical theories Stallo found the most offensively reified, hypostasized, and notional—the most allegorical, daemonic, and empirically unhinged—were actually those most directly in line with the most important coming scientific breakthroughs of the next several decades.

To return to Edward Carpenter's "Modern Science: A Criticism," we can now measure more clearly the discursive disjunctions that fissured that text in its final, 1921 incarnation. Stallo articulated an empirico-skeptical form of rationalist dissent, a hyperpositivist outcry against a heuristic/hypothetical school of physical and mathematical "reification" that produced the seemingly sterile and vacuous absurdities of the ether and hyperspace. It would be hard to locate a philosophical position further removed from the evolutionist and vitalist transcendentalism of Carpenter's own visionary scientisms, with its particular panoply of reified life forces and spiritual exfoliations. But because Stallo made available a sober and plausible attack on the reigning luminaries of theoretical physics, Carpenter was content to lift the former's arguments in pursuit of his own philosophical defenses against the secular presumptions of scientific arrogance. But when Carpenter allowed these cryptopositivist arguments of the 1880s to remain side by side with the new mathematical physics of relativity, non-Euclidean curvatures, and four-dimensional space-time in full swing by the 1920s, the spine of his critique finally broke under the weight of its own anachronistic baggage.

Dimensionality in Zamyatin

In chapter 6 we investigated the primary thermodynamic tonalities of the physical and biological scientisms of Zamyatin's *We*. In that novel the predominance of classical thermodynamical themes mutes the allegory of space and time, but offshoots of this parallel discursive formation occasionally surface in an idiom of alternative dimensionalities. The rhetoric of dimensionality in *We* is compounded largely out of geometrical figures derived from the nineteenth-century merger of ether physics with the mathematical idealisms of the fourth dimension of space, supplemented with the temporal fourth dimension of Einsteinian space-time. The advent of Einstein's theories of relativity as an intellectual and media event of profound cultural as well as scientific import complicates Zamyatin's best-known fictional and critical writings by superimposing mathematical scientisms derived from relativity theory upon prior allegories of spatial dimensionality still bound up with late-classical physical and thermodynamic paradigms. These unresolved historical tensions animate Zamyatin's discursive statements on aesthetics as well as the fictional texture of *We*.

Composed throughout 1920 and completed by the end of 1921, *We* was exactly contemporaneous with the international Einstein boom that followed Eddington's experimental verification of the general theory of relativity. In official Soviet quarters Einstein's ideas were perceived as bourgeois rather than radical, as they ran counter to the entrenched positivist scientisms of Bolshevik orthodoxy.[19] However, in Zamyatin's critical writings, the renovated science of Einstein confronts classical mathematics and physics as the vanguard of an intellectual revolution dividing visionary innovation from plodding dogmatism: "Life itself today has lost its plane reality: it is projected, not along the old fixed points, but along the dynamic coordinates of Einstein, of revolution."[20] Thus the theory of relativity marked a crucial moment of historical demarcation between old and new scientific regimes, and this politico-scientific demarcation precisely parallels the ironic gap dramatized in *We* between OneState's archaic arithmetic and Euclidean orthodoxy and the Mephi rebels' superior mathematical culture.[21]

In addition to their frequent explicit references, Zamyatin's critical essays of the 1920s also acknowledge Einstein's influence by purveying the concept of the fourth dimension in its newly stipulated *temporal* mode. For instance, in "On Synthetism" of 1922, Zamyatin criticized futurist painting for its implicitly Bergsonian imperative of capturing the dynamics of temporal duration through a fixed juxtaposition of images: "One of the Futurists' favorite errors: a hopeless attempt to fix upon the canvas the sequence of several seconds—to introduce into painting, a three-dimensional art, a fourth dimension: continuity, time (a dimension given only to words and to music)."[22] The justice of this remark is certainly debatable, and in its genealogy probably has as much to do with the cubo-futurist fascination with the *spatial* fourth dimension as it does with Einsteinian space-time.[23]

Leaving these polemics aside, however, the temporal fourth dimension makes a more positive appearance in Zamyatin's critical introduction to the work of H. G. Wells: "There are two Wellses: one is an inhabitant of our three-dimensional world, the author of realistic novels; the other is a resident of a world of four dimensions, a traveler in time, the author of tales of science fiction and social fantasy."[24] Zamyatin accords the British writer the particular privilege of having seized the properly temporal or successive axis of verbal narrative to plot his fantastic excursions into new dimensions of science-inspired imaginings. With the triumph of Einstein's revisionary worldview before him, Zamyatin could especially appreciate the predictive power of Wells's prophetic hunch in *The Time Machine,* his purely speculative redefining of the spatial fourth dimension as a reversible direction in time. Wells's prescient turn-of-the-century fantasies thereby earned from Zamyatin the privilege of Einsteinian validation.

A generation after *The Time Machine,* Zamyatin declared in "On Synthetism": "Einstein broke space and time themselves from their anchors. And the art that grows out of this present-day reality—can it be anything but fantastic, dreamlike?"[25] In these terms Zamyatin defends his own and others' forays into non- or "Neo-"realistic literary and artistic forms. Here is a literary-critical scientism captured directly on the cultural battlefield of its place and time: Zamyatin marshaled Einstein's authority to his side in his quixotic conflict with Bolshevik canons of aesthetic propriety. And yet, Zamyatin continued, "even before Einstein," artists and writers had been feeling the tremors of the new physics in the form of the new, hyperspatial geometries: "After the geometrically philosophic earthquake produced by

Einstein, the old space and time perished irrevocably. But even before Einstein, this earthquake was recorded by the seismograph of the new art; even before Einstein, the axonometry of perspective was shaken, the axes of the *X*s, *Y*s, and *Z*s split into multiple rays" (86).

Here Zamyatin refers back to the broad international tableau of non-Euclidean and *n*-dimensional geometries that populated the era of classical thermodynamics, and out of which emerged the particular constructions of a spatial fourth dimension that, in scientistic tandem with the ether of space, warped the cultural field of early modernism long before Einstein's ideas traveled beyond the disciplinary confines of theoretical physics. Zamyatin showed his close familiarity with this prior tradition a year later in "On Literature, Revolution, Entropy, and Other Matters," in the claim that the mid-nineteenth-century Russian mathematician Lobachevsky "cracks the walls of the millennia-old Euclidean world with a single book, opening a path to innumerable non-Euclidean spaces: this is revolution."[26] This essay in particular demonstrates how powerfully and immediately Einstein's Minkowskian space-time geometry became adapted to the rhetorical matrices of previous mathematical formations. In his critical peroration, Zamyatin's geometrical aesthetic dovetailed the non-Euclidean curvatures of Lobachevsky and Riemann with the Minkowskian interpretation of the general theory of relativity:

> Science and art both project the world along certain coordinates. Differences in form are due only to differences in the coordinates. All realistic forms are projections along the fixed, plane coordinates of Euclid's world. These coordinates do not exist in nature. Nor does the finite, fixed world; this world is a convention, an abstraction, an unreality. And therefore Realism—be it "socialist" or "bourgeois"—is unreal. Far closer to reality is projection along speeding, curved surfaces—as in the new mathematics and the new art. (112)

Similarly, Zamyatin's evocations of dimensionality in *We* also present a layering of nineteenth- and twentieth-century, spatial and temporal modes. On the one hand, D-503 appears to possess at least the rudiments of the nineteenth-century hyperspace concepts purveyed in Abbott's *Flatland*. In an early attempt to explain to his unknown audience how difficult it is to see one world from the perspective of another, he writes: "Imagine a square, a splendid, living square. And he has to tell about himself, about his life. You see—the last thing on earth a square would think of telling about is that he has four equal angles. He simply does not see that, it's so familiar to him, such an everyday thing. That's me. I'm in the situation of that square all the time."[27] At this moment in the narrative D-503 is unaware that the revolutionary infiltrator I-330 will soon be playing the role of A. Sphere—the hyperspace visitor and apocalyptic messenger of the three-dimensional millennium—to his A. Square. But despite his nominal sense of dimensional possibilities, D-503's previous indoctrination in the verities of Euclidean geometry and Newtonian calculus have left him ill prepared to comprehend the altered realities thrust upon him as the Mephi plot his seduction.

We immediately establishes the reactionary linearity of D-503's habitual world. Exhorted to use his technological creation—the space rocket INTEGRAL—"to integrate the indefinite equation of the universe," he muses: "Yes: to unbend the wild curve, to straighten it tangentially, asymptotically, to flatten it to an undeviating line. Because the line of OneState is a straight line. The great, divine, precise, wise straight line—the wisest of all lines."[28] Joining in the ranks of the OneState March, he experiences his civic fervor through the beauty of rectilinear geometrical regularities: "I saw everything: the unalterably straight streets . . . the divine parallelepipeds of the transparent dwellings."[29] The obverse of this rectilinear sublime is the moral distress D-503 suffers when the real world does not conform to his mathematical biases. Upon first meeting I-330, D finds that "This woman was just as irritating to me as an irrational term that accidentally creeps into your equation and can't be factored out."[30] He is especially sensitive to furniture arrangements. In an apartment of Ancient House, a relic of the world before OneState, "The lines of the furniture made lopsided ellipses that could never be accommodated in any conceivable equation. I could hardly stand this chaos"; in the poet R-13's room, "He moved one of the easy chairs, then the other, and the planes were dislocated, everything slipped out of the prescribed correlation and became non-Euclidean."[31]

Fixed and regimented temporal and spatial demarcations are proper to the moral calculus and stable geometries that define D-503's sense of OneState and his place in it. The mathematician partitions himself into ego and unconscious, efficient surface and "hairy depth," just as for several centuries the world of OneState has placed itself on one side of the glassy Green Wall that protects it from the hideous chaos of unregimented nature: "O, mighty, divinely delimited wisdom of walls, boundaries! . . . Man ceased to be a wild animal only when he built the first wall. Man ceased to be a wild man only when we built the Green Wall."[32] However, these factitious surfaces begin to crumble as D-503 is drawn into the revolutionary intrigues of femme fatale I-330, and the cracking of these walls is consistently figured by the Lobachevskian path away from Euclidean rectitude. If shadows are the projections of higher-dimensional objects upon lower-dimensional surfaces, I-330's advent within the narrative is "shadowed" by D-503's awareness of being under surveillance, as if by a spy from another dimension: "How come not a day passes that I don't hear his flat feet behind me, squelching along as though walking through puddles? Why is he behind me every day, like a shadow? In front, to the side, behind, this gray-blue, two-dimensional shadow."[33] Once I-330 initiates him into outlaw passion and the sweet torments of sexual jealousy, this guilt dynamic is reinforced by his parallel loss of mass identity: "You're in bad shape," his shadow, the mysterious doctor and Guardian S-4711, tells him: "It looks like you're developing a soul."[34] Implored for an explanation, S-4711 describes the advent of D-503's rebellious individuality in dimensional terms, as a warping of surfaces with the emergence of an interior paraspace: "Take a flat plane, a surface, take this mirror, for instance. And the two of us are on this surface. . . . But just imagine now that some fire has softened this impenetrable surface and nothing skims along the top of it any longer—everything penetrates into it, inside, into that mirror world that we peer into with such curiosity. . . . The plane has taken on mass, body, the world, and it's all inside the mirror, inside you."[35] Thermodynamic intensity and spatial complexity come together in S-4711's clinical allegory of a soul revitalized by an inner heat from another dimension.

At the start of Record 17, D-503 explicitly states the geometrical axes of *We*'s allegorical dimensionality: "new unknowns [have] turned up in my equation," and: "The coordinates of the whole business all began . . . with the Ancient House. The X, Y, and Z axes that recently began serving as the basis of my whole world all start there."[36] D-503's search for a soul "beneath the surface melted by fire" (90) is reinforced by a "mirror-stage" episode of reversed identity. Walking along the darkly translucent Green Wall to the Ancient House, he encounters the "yellow eyes" of an unknown creature: "We looked each other in the eye for a long time—through those shafts connecting the surface world to that other beneath the surface" (91). This blurring of geometric separation is soon compounded by a visionary moment when D views the old babushka who minds the Ancient House: "A yellow stripe of sunlight lay across her knees. And for one second: I, the sun, the old woman, the wormwood, the yellow eyes—we all blended into one, were all bound forever by veins through which flowed one common, stormy, magnificent blood" (91–92). Temporal duration is suspended as the idiom of erotic vitalism erupts within a space of four-dimensional eternity.

This cascading dissolution of boundaries reaches a climax when D, while searching the House for I-330, sights S-4711 close behind him. Recalling a previous episode when I-330 had suddenly disappeared, he steps quickly through the same wardrobe door and stumbles into an unknown dimension.

In this classic moment of allegorical transition from surface to subtext, day-self to dream-self, D-503's model of space is warped again, a new direction in space opens up, as it were a fourth coordinate perpendicular to each of his three prior axes. Like the explorer-narrator of *The Coming Race* encountering the domains of the Ana, or Wells's Time Traveller descending the air-shafts that ventilate the dark realm of the Morlocks, D-503 breaks through the surface of his world and is precipitated into a subterranean limbo. Then I-330 emerges within this seemingly hallucinated paraspace and seduces him once more toward a state of erotic indifferentiation.

As the novel progresses D-503 experiences his new libidinal and existential calculus as a manic-depressive mental breakdown, which Zamyatin continues to narrate under the sign of the hyperdimensionality associated with the revolutionary developments of modern physics and cosmology. D-503 repeatedly uses geometric tropes to embody his feeling that the boundary between dream and waking has collapsed: "Irrational magnitudes are growing up through everything that is stable, customary, three-dimensional, and all around me something rough and shaggy is replacing the firm, polished surfaces."[37] The builder of the INTEGRAL converts his social and erotic anxieties into the hyperbole of hyperbolic geometry, the ideal hypersolids that reside in the spatial fourth dimension:

> For every equation, every formula in the superficial world, there is a corresponding curve or solid. For irrational formulas, for my $\sqrt{-1}$, we know of no corresponding solids, we've never seen them. . . . But that's just the whole horror—that these solids, invisible, exist. They absolutely inescapably must exist. Because in mathematics their eccentric prickly shadows, the irrational formulas, parade in front of our eyes as if they were on a screen. And mathematics and death never make a mistake. And if we don't see these solids on our surface world, there is for them, there inevitably must be, a whole immense world there, beneath the surface. (98)

The rhetorical construction of D-503's epistemological distress submits him to a fate of allegorical oscillation among ontological levels, the protracted vertigo of bouncing in and out of higher and lower dimensions. His old world as well as his old mathematics have become literally *unsprung*. As Brett Cooke puts it: "Indeed, D-503 twice describes himself as a compressed spring: a spiral forced virtually into the shape of a circle."[38] Now, however, despite his efforts to recompose his rational and repetitive circularities, all his customary planar orbits are springing into higher-dimensional spirals. Zamyatin's critical essay "On Synthetism" makes explicit the Hegelian register of the spiral form as a geometrical trope for the dialectical processes of art history that converge to synthesize antithetical formal styles: "A spiral;

a winding staircase in the Tower of Babel; the path of an airplane rising aloft in circles—such is the way of art. The equation of the movement of art is the equation of a spiral."[39] Cooke elicits the implicit dimensional moralization of this figure as one of mathematical liberation from cognitive confinement.[40] And in this direction, D-503's allegorical spirals allude directly to the subtext of earlier fourth-dimensional discourse, where the moving intersection of a spiral with a plane is a common visual metaphor for the complex incursion of higher unto lower dimensions. Within the lower dimension these intersections are cognized as a time series, or as D-503 puts it, as shadows of irrational realities that "parade in front of our eyes as if they were on a screen." In Claude Bragdon's *Primer of Higher Space* of 1913, elaborating an illustration in Charles Howard Hinton's *The Fourth Dimension* of 1904, we read that "the permanent existence of the spiral will be experienced as a time series—the record of traversing the spiral will be a point moving in a circle"[41] (fig. 15).

The subtle interplay of curves and spirals traced in the text of *We* marks the convergence of the old spatial fourth dimension with the new temporal fourth dimension emerging in the discourse of Einstein. Zamyatin's narrative shows again how powerfully the tropes of modern mathematics transferred themselves into images of cognitive and existential possibility. Physics and the mathematics of hyperspace combined throughout the modernist period to drive cultural imaginations of revolutionary transformation. D-503 is both a psychologically plausible character and an allegorically overdetermined signifier of the dislocations caused by the upheavals of mathematical and physical concepts and the corresponding renovations of "depth" in the modern cultural field.

Lawrence Encounters Einstein

D. H. Lawrence's encounter with Albert Einstein was more conceptually successful than his erstwhile mentor's. It occurred not long after Carpenter's publication of the final version of "Modern Science: a Criticism," in Lawrence's *Fantasia of the Unconscious,* completed in the summer of 1921. That June, Lawrence asked a friend to send him a copy of Einstein's popular exposition. Thus the *Fantasia* records Lawrence's feeling for the same early Einstein vogue that drove Carpenter's late revisions and modulated Zamyatin's *We,* as well as Lawrence's own passing acquaintance with some portions of Einstein's brisk and reasonably accessible volume. As a somewhat detached witness to the early ascent of Einstein to cultural stardom,

EVOLUTION INTERPRETED IN TERMS OF HIGHER SPACE

IF WE PASS A HELIX (A SPIRAL IN THREE DIMENSIONS), THROUGH A FILM (A 2-SPACE), THE INTERSECTION WILL GIVE A POINT MOVING IN A CIRCLE [A, FIG 1]. IF THE HELIX REMAINS STILL AND THE FILM MOVES VERTICALLY UPWARD, THE SPIRAL WILL BE REPRESENTED IN THE FILM BY THE CONSECUTIVE POSITIONS OF THE POINT OF INTERSECTION. IN THE FILM THE PERMANENT EXISTENCE OF THE SPIRAL WILL BE EXPERIENCED AS A TIME SERIES—THE RECORD OF TRAVERSING THE SPIRAL WILL BE A POINT MOVING IN A CIRCLE.

FIG 1.

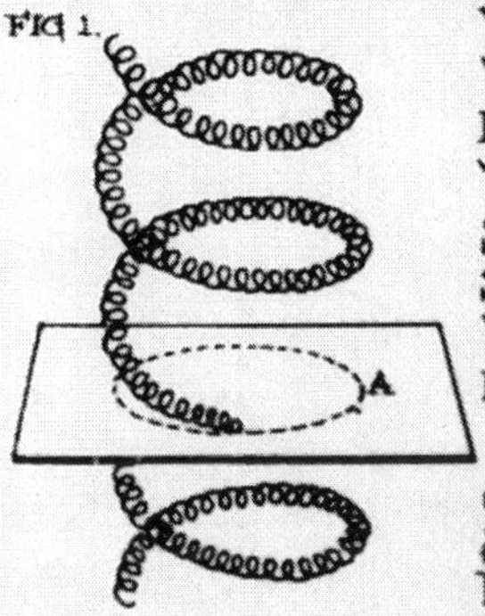

IT IS EASY TO IMAGINE COMPLICATED STRUCTURES OF THE NATURE OF THE SPIRAL, STRUCTURES CONSISTING OF FILAMENTS, AND TO SUPPOSE ALSO THAT THESE STRUCTURES ARE DISTINGUISHABLE FROM EACH OTHER AT EVERY SECTION. IF WE CONSIDER THE INTERSECTIONS OF THESE FILAMENTS WITH THE FILM AS IT PASSES TO BE THE ATOMS CONSTITUTING A FILMAR UNIVERSE, WE SHALL HAVE IN THE FILM A WORLD OF APPARENT MOTION; WE SHALL HAVE BODIES CORRESPONDING TO THE FILAMENTARY STRUCTURE, AND THE POSITIONS OF THESE STRUCTURES WITH REGARD TO ONE ANOTHER WILL GIVE RISE TO BODIES IN THE FILM MOVING AMONGST ONE ANOTHER. THIS MUTUAL MOTION IS APPARENT MERELY. THE REALITY IS OF PERMANENT STRUCTURES STATIONARY, AND ALL THE RELATIVE MOTIONS ARE ACCOUNTED FOR BY ONE STEADY MOVEMENT OF THE FILM.

NOW IMAGINE A <u>FOUR</u>-DIMENSIONAL SPIRAL PASSING THROUGH A <u>THREE</u>-DIMENSIONAL SPACE. THE POINT OF INTERSECTION, INSTEAD OF MOVING IN A CIRCLE, WILL TRACE OUT A SPHERE. ASSUMING, AS BEFORE, A COMPLICATED STRUCTURE FOR THE SPIRAL, ITS PRESENTMENT IN 3-SPACE WILL CONSIST OF BODIES BUILT UP OF SPHERES OF VARIOUS MAGNITUDES MOVING HARMONIOUSLY AMONGST ONE ANOTHER, AND REQUIRING <u>TIME</u> FOR THEIR DEVELOPMENT. MAY NOT THE ATOM, THE MOLECULE, THE CELL,—THE EARTH ITSELF BE SO MANY PATHS AND PATTERNS OF AN UNCHANGING UNITY?

PLATE 16

Fig. 15. Evolution Interpreted in Terms of Higher Space. From Claude Bragdon, *A Primer of Higher Space* (1913).

Lawrence noted the odd glamor of his famously recondite theory: "The latest craze is Mr. Einstein's Relativity Theory. Curious that everybody catches fire at the word Relativity. There must be something in the mere suggestion, which we have been waiting for. But what?"[42] Beyond the specifically scientific meanings that the theory of relativity proposes with regard to the physical world, Lawrence was concerned to grasp the cultural meanings of its intellectual vogue.[43]

Lawrence first discussed Einstein at the beginning of the second chapter of the *Fantasia:*

> We are all very pleased with Mr. Einstein for knocking that eternal axis out of the universe. The universe isn't a spinning wheel. It is a cloud of bees flying and veering around. Thank goodness for that, for we were getting drunk on the spinning wheel. So that now the universe has escaped from the pin which was pushed through it, like an impaled fly vainly buzzing: now that the multiple universe flies its own complicated course quite free, and hasn't got any hub, we can hope also to escape.[44]

This bridge mix of metaphors alludes to one of the primary effects of relativity theory on the classical cosmos, the elimination from physical consideration of an absolutely stationary frame of cosmic reference—an ultimate system of Cartesian coordinates or absolute set of x-, y-, and z-axes—by which one could then absolutely determine the position of objects in space and the velocities of objects in motion through space. The "pin" that Lawrence credited Einstein with pulling out of the universe is probably the ether as absolute frame of reference. Whereas before, physics posited a single standard of absolute position and motion, now there is truly is a "multiple universe" divisible into an infinite set of reference frames.

Lawrence, of course, was not primarily interested in the physics of space and time. His most concerted attempt to summarize the theory of relativity tended to homogenize Einstein's physical themes into equivocal matters of "force," and without a clear appreciation of the geometric or spatiotemporal implications of the theory. "As far as I can see, Relativity means, for the common amateur mind, that there is no one absolute force in the physical universe, to which all the other forces may be referred. There is no one single absolute central principle governing the world. The great cosmic forces or mechanical principles can only be known in their relation to one another, and can only exist in their relation to one another."[45] For Lawrence, rather, the most important implication of relativistic relationality was its availability as an authoritative metaphor with which to express, paradoxically, the metaphysics of individuality.

"There is only one clue to the universe," Lawrence asserted the *Fantasia:*

"And that is the individual soul within the individual being. That outer universe of suns and moons and atoms is a secondary affair."[46] He welcomed the theory of relativity insofar as it could be plausibly extended to matters of selfhood and its proper liberation from fixed and/or collective schemes of identity. Lawrence was at least ironically willing to abet a scientistic extension of relativity as long as it could be coordinated with his own philosophical agenda for an ethics of transcendental subjectivity. Just like the new Einsteinian universe, Lawrence declared, "we won't be pinned down, either. We have no one law that governs us. For me there is only one law: *I am I*. And that isn't a law, it's just a remark. One is one, but one is not all alone. There are other stars buzzing in the center of their own isolation. And there is no straight path between them."[47] Here Lawrence bent to his own discursive purpose the specific component of the general theory of relativity verified so dramatically by the measurements taken during the solar eclipse of 1919. In Einstein's phrasing: "*in general, rays of light are propagated curvilinearly in gravitational fields.*"[48] Lawrence twisted this famous Einsteinian nonlinearity further in cajoling his reluctant reader to "apply a little theory of relativity, and realize that what I say is not what you hear, but something uttered in the midst of my isolation, and arriving strangely changed and travel-worn down the long curve of your own individual circumambient atmosphere."[49] Lawrence's playful scientism implies that we are each of us a separate world in motion through deep space. Anticipating the critical relativity of contemporary literary theory, he notes that the reception of human communications can deviate from their transmission as sharply as light beams bent out of rectitude by the gravity of solar masses.

After this early and fairly solid invocation of relativity, Lawrence dismissed Einstein for much of the book. But toward the end he reappears in two more passages. In chapter 13, titled "Cosmological," Lawrence took from Einstein's challenges to classical mechanics a conviction that "the universe is once more in the mental melting-pot."[50] Einstein was bringing about a cosmological shake-up broad enough to include his own mythopoetic revisions of psychological dynamics. At the same time, Lawrence cautioned against reading relativity as the announcement of a total revolution in cosmology. For one, "Mr. Einstein's Theory of Relativity does not supersede the Newtonian Law of Gravitation or of Inertia. It only says, 'Beware! The Law of Inertia is not the simple ideal proposition you would like to make of it. It is a vast complexity. Gravitation is not one elemental uncouth force. It is a strange, infinitely complex, subtle aggregate of forces.' And yet, however much it may waggle, a stone does fall to earth if you drop it" (180). Lawrence's basic reservation about relativistic absolutism and his celebration of cosmic complexity are sound; however, relativity theories do not simply make simple Newtonian propositions more complex, but rather provide a new universal framework—specifically, a non-Euclidean geometry—for unifying the classical dynamics of matter in motion with the new physics of gravitational and electromagnetic fields. That is, in important ways the theory of relativity really does supersede the Newtonian system, whose range of practical application is still impressive but is now seen to be restricted to cases of relatively low velocities and weak fields. The aim of Lawrence's delimitation of the physical theory, however, is most importantly to provide a scientistic preparation for his own radical ethical conservatism: "We should like, vulgarly, to rejoice and say that the new Theory of Relativity releases us from the old obligation of centrality. It does no such thing. It only makes the old centrality much more strange, subtle, complex, and vital. It only robs us of the nice old ideal simplicity" (180).

What then is this "old obligation of centrality" that the theory of relativity appears to make more complex? It concerns the ultimate term and prime conserved principle in Lawrence's psychological cosmology, namely the continued absolutism of the individual self. Lawrence articulates this moral axiom in his final mention of the theory of relativity. He reiterates his overall sympathy with what he understands to be the gist of Einstein's message: "I feel inclined to Relativity myself. I think there is no one absolute principle in the universe. I think everything is relative. But I also feel, most strongly, that in itself each individual living creature is absolute: in its own being. And that all things in the universe are just relative to the individual living creature. And that individual living creatures are relative to each other."[51] Like most of us, Lawrence wanted it both ways: relativity when it helps, the absolute the rest of the time.

In the encounter between Lawrence and Einstein, then, Lawrence's own visionary scientisms did not challenge science in its entirety. Rather, they used Einsteinian relativity to attack the cultural powers of classical positivism and determinism, those late-Newtonian scientisms that presumed to order human values solely according to the impersonal calculation of nonhuman forces. In the *Fantasia,* Lawrence located his satisfaction with the latest physical challenges to classical science not on the basis of any precise comprehension of their specifics, but on the *affect* conveyed by the mutable and indeterminate phenomena they discover. Lawrence did not make it fully across the conceptual threshold from classical absolutism into Einstein's relativized spatiotemporal world. Rather, in a discursive move reminiscent of Ursula Brangwen's response to Dr. Frankstone in *The Rainbow,* he redirected the argument away from the sheer impersonality of nonvital scientific entities—physical masses, velocities, forces, and fields—and by some relatively unsubtle equivocations, wrenched the notion of relativity out of mathematical physics and into his system of vitalistic ethics. The result, again as in Ursula's case, was the construction of a visionary scientism, in which the aura of authoritative scientific cachet is conducted, however antithetically, to a counterscientific enterprise. As is also the case with Ursula, Lawrence's scientisms had recourse to relatively outmoded forms of superannuated science in order to secure a special position, a point of existential reference that transcends the equality that is the actual physical effect of Einstein's theories of relativity. But despite these literal deviations from contemporary conceptual rectitude, Lawrence's extended metaphors and mythopoetic flights can still shake up the cultural meanings we get out of recondite scientific discourses. As Lawrence asserted when he first confronted Einstein early in the *Fantasia,* "we are in sad need of a theory of human relativity. We need it much more than the universe does."[52]

Lawrence's Fourth Dimension

Lawrence's encounter with Einstein had a subtle but significant effect on his writings after 1921, an aftereffect that provides a measure of his concern to update his own critical doctrines by continued conversation with current scientific developments. The situation is compounded, however, by Lawrence's further encounter in 1924 with P. D. Ouspensky's *Tertium Organum: A Key to the Enigmas of the World.*[53] The physicist Einstein and the philosopher Ouspensky converged in one salient particular. Both offered a cosmological discourse of the "fourth dimension" along with a critique of the concept of the ether. As we have noted in the two previous chapters, in the period when Lawrence composed the text of the *Fantasia* and specifically addressed Einstein's theories of relativity, the ether of space remained an important fixture in his writings. It is probable, however, that his acquaintance with the theory of relativity initiated Lawrence's awareness that the spatial fixity of the ether-concept was in fact the hypothetical "pin" that Einstein had pulled out of the cosmos.

Ouspensky's own consignment of the ether to an outmoded phase of "three-dimensional" physics may have provided a further impetus for Lawrence to reconceive his scientistic figures. In a series of significant doctrinal essays revised or composed in 1925 during or just after Lawrence's final sojourn in New Mexico, we find evidence that the new physical currency Einstein gave to the concept of the fourth dimension, along with the renewed philosophical currency provided simultaneously by *Tertium Organum,* moved Lawrence to revise his own applications of the ether concept. In Lawrence's subsequent writings of that year, the fourth dimension emerges explicitly, manifesting a certain Einsteinian warrant to broach conceptions that are in fact much closer to Ouspensky's timely reprise of Charles Howard Hinton. At the same time, the Einstein vogue also gave Lawrence an authoritative physical standpoint from which to counter Ouspensky's particular brand of idealist metaphysics.

Before 1924–25, Lawrence had brought the ether forward in two different tonalities. Now and then he treated it in its authorized scientific role as an undemonstrated but theoretically necessary physical medium for the propagation of radiant energies. At these times, Lawrence often echoed both the phrasings and the scoffing tone of Edward Carpenter's problematic adaptation of J. B. Stallo in "Modern Science: A Criticism." In the *Fantasia,* following upon his observation that Einstein had placed "the universe . . . once more in the mental melting-pot," Lawrence paralleled Carpenter in connecting the atom and the ether as fundamentally factitious entities: "The atom? Why, the moment you discover the atom it will explode under your nose. The moment you discover the ether it will evaporate. The moment you get down to the real basis of anything, it will dissolve into a thousand problematic constituents."[54] Lawrence's antifoundational intuitions about the limits of mechanistic reductionism remain well taken, but the discursive logic of his argument derives from Stallo's classical positivism more than from post-Einsteinian relativism or quantum-theoretical indeterminism. The same basic litany had been put back into play in Lawrence's 1925 revisions to *The Crown:* "These 'laws' which science has invented . . . and even these absolute facts, like—the earth goes round the sun—or the doubtful atoms, electrons, or ether—they are all prison-walls, unless we realise that we don't know what they mean."[55] In passages like these, Lawrence joined Carpenter and Ouspensky in celebrating the demise of the ether-concept, long predicted by positivist skeptics on empiricist grounds, as a fitting rebuke to the overweening deterministic scientism of classical physics. Unlike either Einstein or Ouspensky, however, Lawrence would not publicly credit science as a progressive enterprise, for which the lapse of the ether is at least an informative episode in a major paradigm shift, if not a momentous movement into a more coherent scientific depiction of the cosmos. Eliding these possible readings, Lawrence mostly dismissed the ether from conceptual and rhetorical duty, in favor of an explicit discourse of the fourth dimension. In doing so he transferred the attributes of divinity he most cherished—creation and vitality—to the fourth dimension, while eliminating them from the ether.

At least from the later 1910s through the mid-1920s, Lawrence had embraced the ether in its unauthorized transcendental role as a medium of vitalistic energies and spiritual communications, and in most of these instances his treatment was rhapsodic rather than sardonic. Abruptly in 1925, however, the idiom of the ether is phased out of Lawrence's text. To my knowledge, its last significant appearance occurs in "Why the Novel Matters," probably written late in 1925, after the main emergence of the fourth dimension as a discursive trope. In a late reprise and critique of "all this spirit stuff"—presumably theosophical ideas that had once influenced Lawrence's theologized ether idiom—he vacated the attribute of mystic vitality from the ether and returned that medium to the physical and technological registers:

> Every man, philosopher included, ends in his own finger-tips. That's the end of his man alive. As for the words and thoughts and sighs and aspirations that fly from him, they are so many tremulations in the ether, and not alive at all. But if the tremulations reach another man alive, he may receive them into his life. . . . The so-called spirit, the message or teaching of the philosopher or the saint, isn't alive at all, but just a tremulation upon the ether, like a radio message. All this spirit stuff is just tremulations upon the ether. If you, as man alive, quiver from the tremulations of the ether into new life, that is because you are man alive, and you take sustenance and stimulation into your alive man in a myriad ways.[56]

Lawrence displaced the spiritual functions of his earlier ether concept to an equally idiosyncratic construction of the fourth dimension. His brief encounter with *Tertium Organum* in 1924 seems to have reminded him that just as the metaphysical ether had developed alongside the now discarded scientific form of the luminiferous ether, so too there was a metaphysical fourth dimension that anticipated Einstein's application of Minkowski's four-dimensional space-time continuum. In addition, his sudden invocations of van Gogh and Cézanne in these same essays may indicate Lawrence's renewed awareness that an aesthetic discourse on the spatial fourth dimension had also flourished in important theoretical discussions of postimpressionist, cubist, and futurist art. In reacting to Ouspensky's elaborate reprise of the spatial and evolutionist fourth dimension of hyperspace philosophy, in fact, Lawrence located a way to carry the mystic or absolute ether forward into the new era of relativity and space-time. Lawrence would defend and justify the creation and reception of *art* as an affair profoundly connected to "the much-debated fourth dimension."[57]

The distant, perhaps multiply refracted echo of Hinton in Lawrence's declaration to Mabel Dodge Luhan—"One's got to put a new ripple in the ether"—suggests that Lawrence was as familiar with the earlier, pre-Ouspenskian discourses of hyperspace that developed in the decades leading up to Einstein as he was with the closely related metaphysical discourses of the ether. For instance, he could have encountered the fourth dimension as early as 1905 at the beginning of Carpenter's *The Art of Creation,* where the notion is articulated in affirming a philosophical and methodological axiom that Lawrence would also adhere to: "The problem of the Universe is essentially insoluble without the introduction of some transcendent factor—Will, Being, the Ego, a Fourth dimension, or whatever we may term it."[58] Richard O. Young has determined that Lawrence was familiar with some form of the idea, "probably not for the first time, in about 1907 when he and Jessie Chambers read Reginald J. Campbell's *The New Theology.* To Campbell, the fourth dimension was Heaven and Jesus was a fourth-dimensional being."[59] In earlier sources such as these, Lawrence would have encountered the idea of the fourth dimension in a spiritualized configuration parallel to the spiritualized ether of space. By 1908 he had gathered the idea of the fourth dimension to his youthful apprehension of a vitalistic cosmology.[60] In addition, from 1912 on the discussion of modern painting in relation to the fourth dimension of space was available in the seminal manifestoes of cubism and futurism.[61]

Like many earlier studies of the fourth dimension in the modernist period, Young's broader cultural chronology is constricted by the anachronistic conflation of the four-dimensional space-time continuum of general relativity with the spatial fourth dimensions of *The Unseen Universe,* Hinton's *Scientific Romances,* and Abbott's *Flatland.* Nevertheless, Young's essay convincingly indicates that a rhetoric of dimensionality operated throughout Lawrence's writings, especially in the form of a climactic crossing or transgression of dimensional thresholds—the way that so many of Lawrence's questing or trespassing characters "break out" of confining spaces, "hop off" one plane of existence in favor of another, or discover a "new heaven and earth" in some parallel dimension that opens off of the known world in some unknown but discoverable direction. Dimensional analogies are without question a basic mode of Lawrencean allegory. Young's textual samples also indicate that Lawrence's implicitly fourth-dimensional metaphors are consistently *spatial* in construction, and thus, we would add, pre-Einsteinian in formation.

However, an explicit rhetoric of the fourth dimension did not actually emerge in Lawrence's text until late in 1924, and then again in the 1925 revision or composition of essays collected in *Reflections on the Death of a Porcupine.* Part of Young's chronological confusion stems from the fact that Lawrence himself not only understated the extent of the revisions he made to *The Crown* in 1925—"I alter *The Crown* only very little"—but also encouraged their backdating to the earlier version "written in 1915."[62] The textual apparatus of the 1988 Cambridge edition lets one reconstruct the actual chronology of *The Crown*'s composition and thus discover that many of the text's most striking formulations—for instance, the sustained diatribe against the ego—are late interpolations. Although neither the 1915 nor the 1925 *Crown* names the fourth dimension explicitly, its conceptual traces are clearly implicated in the following revision. Whereas the 1915 version had read, "In Time and in Eternity all is flux. Only in timelessness is Heaven" (478), the 1925 text reads: "In Time and in Eternity all is flux. Only in the other dimension, which is not the time-space dimension, is there Heaven. We can no more *stay* in this heaven, than the flower can stay on its stem. We come and go" (304). Lawrence's phrase of 1925 "time-space dimension" reworks Einstein's standard locution of the "four-dimensional space-time continuum," while at the same time paralleling Ouspensky's demotion of Einsteinian space-time in favor of a conception of the fourth dimension from which time has been annihilated. "The other dimension, which is *not* the time-space dimension" offers Lawrence a newly scientized circumlocution of the philosophical notion of timelessness per se—the "mystic Now" in Lawrence's poetics of momentaneity.

The essential spatiality and corresponding atemporality of Lawrence's fourth dimension is particularly clear in his doctrinal essay "Reflections on the Death of a Porcupine," which energetically formulates the distinction between mere *existence* and full *being* as one between being located in the "time-space dimension" where relativity holds, or in its antithesis, the "fourth dimension" of the absolute self:

1. Any creature that attains to its own fullness of being, its own *living* self, becomes unique, a nonpareil. It has its place in the fourth dimension, the heaven of existence, and there it is perfect, it is beyond comparison.
2. At the same time, every creature exists in time and space. And in time and space it exists relatively to all other existence, and can never be absolved.[63]

Here Lawrence has adapted the uncanny geometry of the spatial fourth dimension to his vision of momentaneous consummation by which the body and soul together may bypass the world of mechanical determinism by a transcendental spatial movement perpendicular to all known directions. We return in some fashion to the rectilinear but substantial hyperspace of Hinton's imagination, into and out of which our four-dimensional bodies may "come and go": "Being is *not* ideal, as Plato would have it: nor spiritual. It is a transcendent form of existence, and as much material as existence is. Only the matter suddenly enters the fourth dimension" (359). And yet Lawrence also keeps in operation a non-Euclidean or Einsteinian vocabulary of curved space, when he further specifies his fourth dimension as a "mysterious other reality of things in a perfected relationship. It is into this perfected relationship that every straight line curves, as if to some core, passing out of the time-space dimension" (361).

Lawrence's fourth dimension is an uncanny paraspace one enters by passing out of time, momentarily, through successfully placing oneself into a "perfected relationship" with an other, another being or self. It is also an Imaginary space that sutures the gaps of the material and bodily Real. His spatial fourth dimension is a continuous void or empty plenum into which the life force of evolutionary vitalism and hence, creative desire, thrusts itself as it crests over from mere existence into full being. In "Morality and the Novel," Lawrence improvises a passage of art criticism in order to restate this point:

> When Van Gogh paints sunflowers, he reveals, or achieves, the vivid relation between himself, as man, and the sunflower, as sunflower, at that quick moment of time. . . . You cannot weigh nor measure nor even describe the vision on the canvas. It exists, to tell the truth, only in the much-debated fourth dimension. In dimensional space it has no existence.
>
> It is a revelation of the perfected relation, at a certain moment, between a man and a sunflower. It is neither man-in-the-mirror nor flower-in-the-mirror, neither is it above or below or across anything. It is in-between everything, in the fourth dimension.[64]

In this latter locution Lawrence compresses the infinitesimal infinitude of the spatial fourth dimension, its absolute "in-between-ness" or quality of leading off from any given point in an unknown direction. Lawrence's locution recalls Hinton's theory of the ether as the surface where the third dimension rests on the fourth. Or more nearly, it echoes Marcel Duchamp's notion of the "infrathin," which Linda Henderson has expounded as "the qualitative degree or step by which one form turns into another, a quality he found at the critical point of transition from line to plane to solid and beyond."[65] Duchamp made the following note concerning this notion: "allegory / (in general) / is an application / of the infra thin."[66] The casual rigor of Lawrence's allegorical treatments of four-dimensional spatiality make it difficult to imagine that he was not familiar with the idioms of art criticism that circulated in the cosmopolitan groups around Gertrude Stein and Mabel Dodge Luhan. In *Les Peintures Cubistes,* for instance, Guillaume Apollinaire wrote in 1913: "As it presents itself to the mind, from the plastic point of view, the fourth dimension appears to be engendered by the three known dimensions: it represents the immensity of space eternalizing itself in all directions at any given moment. It is space itself, the dimension of the infinite."[67]

In the period following his leadership novels *Aaron's Rod, Kangaroo,* and *The Plumed Serpent,* Lawrence redefined many of his teleological schemes in ways that removed both the urgency of direct political organization and the more strident advocacy of male primacy in forging new sociopolitical as well as erotic directions. Where now would Lawrence lead his reader? It is a question, simply stated, of finding a "new direction" within human space and time. Lawrence tackled this problem explicitly in "Him with His Tail in His Mouth," a doctrinal satire on the Ouroboros or serpent that bites its own tail as a symbol of infinity and eternity. For Lawrence the Ouroboros symbolizes the way that most traditional spiritual guidebooks giving existential directions lead one in the end to seek a return to the origin. Lawrence rebutted these theocentric and logocentric pretensions of "religion and philosophy," which have declared not only that the beginning and the end are one and the same, but also "that the beginning of all things is the life-stream itself, energy, ether, libido, not to mention the Sanskrit joys of Purusha, Pradhana, Kala."[68] That is, if the ultimate cause responsible for the unfolding of the human or material universe in just the way we find it can be specified, then we can then direct our efforts toward the end of knowing and merging with this beginning. And in that case, we seem to know which way to go, what direction our spiritual lives should take.

From the syncretic Eastern gods of Western theosophy, Bergson's élan vital, and the Freudo-Jungian libido to the energy laws and ether theories of Victorian physicists, Lawrence castigates both science and religion for the spiritual bullying performed by these exaggerated claims to dictate the proper direction to the possession of ultimate knowledge. In his late hymn to Cézanne, *Introduction to These Paintings,* Lawrence decried science in particular for its imperious and impersonal reductiveness: "so many modern scientists . . . work with the mind alone, and *force* the intuitions and instincts into a prostituted acquiescence."[69] He offered instead a kind of unknowing knowledge for which no standardized itinerary is possible. For although "there is such a thing as life, or life-energy . . . where it comes from, or what it is, we don't know, and never shall."[70] Lawrence's high vitalist faith in the abiding mystery of Life *in its perfected fullness*—the permanent escape of *being* from our various explanatory schemes—finesses the scientific issue by removing Life to a position beyond mere physical and biological *existence.* However, if we cannot know the beginning or the end, how is one direction any better than another? If Life has neither origin nor end, how can we move in its direction? His answer is—by trusting it enough to turn away from it. The unknowable dark river of Life "flows into us from behind and below. We must turn our backs to it, and go ahead. . . . But which way is ahead?"[71]

The fourth dimension enters the text of "Him With His Tail in His Mouth" at this point to provide a complex metaphor for the direction living things must go in order to have Life drive them onward into Being. Rejecting Darwinian accounts, in what Lawrence jocularly refers to as "the true history of evolution," he reconceives the origin of diverse species as the taking of separate and "distinct roads across the fourth dimension":[72] "When the cuckoo, the cow, and the coffee-plant chipped the Mundane Egg, at various points, they stepped out, and immediately set off in different directions. Not different directions of space and time, but different directions in creation: within the fourth dimension" (312–13). Rewriting and synthesizing previous evolutionary vitalisms, Lawrence amalgamated and reworked both the Bergsonian and hyperspatial schemes, retaining some the unpredictable irreversibility of Bergson's creative evolution while modulating it toward the metatemporal geometries of hyperspace discourse: "As for the goal, which doesn't exist, but which we are always coming back to: well, it doesn't spatially, or temporally, or eternally exist: but in the fourth dimension, it does" (314).

Lawrence's fourth dimension also brings about the rewriting of his earlier rhetoric of equilibrium, as in the "star equilibrium" Birkin had offered Ursula in *Women in Love,* or in his many tropes constructed on the "balance of polarities." To some extent these later essays deflect the more salient technoscientific resonances of thermodynamic equilibrium and electromagnetic polarity by attributing his conceptions to the pre-Socratics: "I want, in the Greek sense, an equilibrium between me and the rest of the universe. . . . Equilibrium, in its very best sense—in the sense the Greeks *originally* meant it—stands for the strange spark that flies between two creatures, two things that are equilibrated, or in living relationship. It is a goal: to come to that state when the spark will fly."[73] But the luminiferous and electromagnetic ether remains embedded in this description of the fourth dimension as a medium for the propagation and reception of vital "sparks," entities defined in physics as momentary luminous disruptive electrical discharges. Lawrence's fourth dimension is a dynamical system or rhetorical energy field within which he can revise previous discursive and fictive tropes both of temporal process, as in the irreversible and "exquisite" trembling of the existential balance, and of the instantaneous sparks that "fly" through space when the balance momentarily pauses at a point of perfected equilibrium: "How marvellous is . . . the exquisite frail moment of pure conjunction, which, in the fourth dimension, is timeless" (316).

The metamorphoses of the electromagnetic field that began with the Romantic unfolding of physical polarities, that developed through the force fields of Faraday and the luminiferous ether media of Maxwell and his circle, that accompanied a geometrical revolution climaxed by the local and universal relativities of Einstein, have arrived in Lawrence at a moment finely poised between doctrinal obsolescence and mythopoetic stability. In this canny critical revision of modernist technoscientisms, the fourth dimensions of philosophical hyperspace, evolutionary vitalism, and mathematical physics pass by an axis perpendicular to them all into the fourth dimension of Lawrencean art-speech. In this existential allegory of space and time, modernism's fourth dimensions take up textual cohabitation in a complex trope of selfhood as an absolute relativity founded on the propagation of "frail moments of pure conjunction." Along with their ostensible organicism and vitalist commitments, however, Lawrence's allegories of ethereal and four-dimensional space and momentaneous time are also sociospiritual ripples of the development of electronic communication technology out of the ferment of energy forms in the era of classical thermodynamics.

Notes

Introduction

1. N. Katherine Hayles, *Chaos Bound: Orderly Disorder in Contemporary Literature and Science* (Ithaca, N.Y.: Cornell University Press, 1990), 8.

2. In *The Last Three Minutes* (London: Weidenfeld and Nicholson, 1994), Paul Davies credits the author of the concept of *Wärmetod* in 1856, Hermann von Helmholtz, with

> what is probably the most depressing prediction in the history of science. The universe, Helmholtz claimed, is dying. The basis of this apocalyptic pronouncement was the so-called second law of thermodynamics. . . . The work of Helmholtz, Rudolf Clausius, and Lord Kelvin led to a recognition of the significance of a quantity called entropy for characterizing irreversible change in thermodynamics. . . . The discovery that the universe was dying as an inexorable consequence of the laws of thermodynamics had a profoundly depressing effect on generations of scientists and philosophers. (9–12)

3. William Thomson, "On a Universal Tendency in Nature to the Dissipation of Mechanical Energy," in *Mathematical and Physical Papers* (Cambridge: Cambridge University Press, 1911), 1:514 (first published in *Philosophical Magazine,* October 1852).

4. Gillian Beer, "Wave Theory and the Rise of Literary Modernism," in *Realism and Representation: Essays on the Problem of Realism in Relation to Science, Literature, and Culture,* ed. George Levine (Madison: University of Wisconsin Press, 1993), 196.

5. Linda Dalrymple Henderson, *The Fourth Dimension and Non-Euclidean Geometry in Modern Art* (Princeton: Princeton University Press, 1983); see also her *Duchamp in Context: Science and Technology in the "Large Glass" and Related Works* (Princeton: Princeton University Press, 1998).

6. See Anson Rabinbach, *The Human Motor: Energy, Fatigue, and the Origins of Modernity* (New York: Basic Books, 1990).

7. A sampling of the recent literature includes Henry Krips, J. E. McGuire, and Trevor Melia, eds., *Science, Reason, and Rhetoric* (Pittsburgh: University of Pittsburgh Press, 1995); Mary Hesse, "Models, Metaphors, and Truth," in *From a Metaphorical Point of View: A Multidisciplinary Approach to the Cognitive Content of Metaphor,* ed. Zdravko Radman (Berlin: de Gruyter, 1995), 351–72; Mary Ellen

Pitts, "Reflective Scientists and the Critique of Mechanistic Metaphor," in *The Literature of Science: Perspectives on Popular Scientific Writings,* ed. Murdo William McCrae (Athens: University of Georgia Press, 1993), 249–72; James J. Bono, "Science, Discourse, and Literature: The Role/Rule of Metaphor in Science," in *Literature and Science: Theory and Practice,* ed. Stuart Peterfreund (Boston: Northeastern University Press, 1990), 59–89; Eric Zencey, "Entropy as Root-Metaphor," in *Beyond the Two Cultures: Essays on Science, Technology, and Literature,* ed. Joseph W. Slade and Judith Yaross Lee (Ames: Iowa State University Press, 1990), 185–200; Darko Suvin, "Science-Fiction: Metaphor, Parable, and Chronotope," in *Just the Other Day: Essays on the Suture of the Future,* ed. Luk de Vos (Antwerp: EXA, 1985), 81–99.

8. On contemporary allegory theory in the context of the tradition of literary metamorphosis, see my *Allegories of Writing: The Subject of Metamorphosis* (Albany: State University of New York Press, 1995).

9. Walter Benjamin, *The Origin of German Tragic Drama,* trans. John Osborne (New York: Verso, 1985).

10. On the link between the baroque emblem and the modern commodity, see Terry Eagleton, *Walter Benjamin; or, Towards a Revolutionary Criticism* (London: NLB, 1981): "the commodity can be understood as the baroque emblem pressed to an extreme" (30).

11. Susan Buck-Morss, *The Dialectics of Seeing: Walter Benjamin and the Arcades Project* (Cambridge: MIT Press, 1988).

12. Bruno Latour, *Science in Action: How to Follow Scientists and Engineers through Society* (Cambridge: Harvard University Press, 1987).

13. I discuss Carpenter and Marsden further in *Dora Marsden and Early Modernism: Gender, Individualism, Science* (Ann Arbor: University of Michigan Press, 1996).

14. Betty Jo Teeter Dobbs and Margaret C. Jacobs, *Newton and the Culture of Newtonianism* (Atlantic Highlands, N.J.: Humanities Press, 1995).

15. See Ilya Prigogine and Isabelle Stengers, *Order Out of Chaos: Man's New Dialogue with Nature* (New York: Bantam, 1984); and Peter Coveney and Roger Highfield, *Frontiers of Complexity: The Search for Order in a Chaotic World* (New York: Fawcett Columbine, 1995).

16. See, for instance, Ian F. A. Bell, *Critic as Scientist: The Modernist Poetics of Ezra Pound* (New York: Methuen, 1981); Michael Wutz, "The Energetics of *Tarr:* The Vortex-Machine Kreisler," *Modern Fiction Studies,* 38, no. 4 (1992): 845–69; Clarke, "Vitalism, Evolution, and Thermodynamics," in *Dora Marsden,* 26–32; Daniel Albright, *Quantum Poetics: Yeats, Pound, Eliot, and the Science of Modernism* (Cambridge: Cambridge University Press, 1997); and Henderson, *Duchamp in Context,* 98–120 and passim.

17. Erik Davis, *Techgnosis: Myth, Magic, and Mysticism in the Age of Information* (New York: Harmony Books, 1998); and Friedrich A. Kittler, *Gramophone, Film, Typewriter,* trans. Geoffrey Winthrop-Young and Michael Wutz (Stanford, Calif.: Stanford University Press, 1999).

Chapter 1

1. "Fire finishes off the horses, strikes them down. The source, the origin, of force is in this flash of lightning, this ignition. Its energy exceeds form; it transforms. Geometry disintegrates, lines are erased: matter, ablaze, explodes. . . . Here comes Turner. . . . The first true genius in thermodynamics." Michel Serres, "Turner Translates Carnot," in *Hermes: Literature, Science, Philosophy,* ed. Josué V. Harari and David F. Bell (Baltimore: Johns Hopkins University Press, 1982), 56–57.

2. The literature on allegory and its theory is vast, but its current discussion as a literary topic is most indebted to the work of Walter Benjamin, in particular, *German Tragic Drama;* Paul de Man, *Allegories of Reading: Figural Language in Rousseau, Nietzsche, Rilke, and Proust* (New Haven: Yale University Press, 1979); and Angus Fletcher, *Allegory: The Theory of a Symbolic Mode* (Ithaca, N.Y.: Cornell University Press, 1964). In my *Allegories of Writing,* I discuss all three and provide numerous further references to the critical literature on this topic. Recent works of note on literary allegory are Deborah L. Madsen, *Rereading Allegory: A Narrative Approach to Genre* (New York: St. Martin's, 1994); Gordon Teskey, *Allegory and Violence* (Ithaca, N.Y.: Cornell University Press, 1996); and Theresa M. Kelley, *Reinventing Allegory* (Cambridge: Cambridge University Press, 1997). *Energy Forms* is the first book-length effort to develop the relation between allegory and science.

3. "By a model is meant a mathematical construct which, with the addition of certain verbal interpretations, describes observed phenomena," John von Neumann, "Method in the Physical Sciences," in *Collected Works,* ed. A. H. Taub, 6 vols. (New York: Macmillan, 1963), 6:492. D. W. Theobald explains that

> the model in terms of which we consider a physical system will decide for us *how* our theory is to be compared with observation. . . . A model in fact determines the domain of application of a formalism by suggesting appropriate interpretation rules. These rules enable us to develop the implications of theoretical concepts to the point at which a comparison between theory and observation is possible. A *comparison* of theory and observation is *all* that can be achieved. . . . Models then are . . . the physical conceptions we have of the structure of any physical system.

D. W. Theobald, *The Concept of Energy* (London: E. and F. N. Spon, 1966), 17–18.

4. See Crosbie Smith, *The Science of Energy: A Cultural History of Energy Physics in Victorian Britain* (Chicago: University of Chicago Press, 1998).

5. James Clerk Maxwell to Michael Faraday, November 9, 1857, in *The Scientific Letters and Papers of James Clerk Maxwell,* ed. P. M. Harman, 2 vols. (Cambridge: Cambridge University Press, 1990), 1:549.

6. James Clerk Maxwell, "Report on Tait's Lecture on Force:—B.A., 1876," in Lewis Campbell and William Garnett, *The Life of James Clerk Maxwell* (London: Macmillan, 1882), 646–48.

7. "Mechanics suffered from a considerable confusion of nomenclature which persisted well into the nineteenth century. The word 'force' was often used indis-

criminately to refer to both the Newtonian concept F and the effect of this force acting through distance, that is $\frac{1}{2}mv^2$" (Theobald, *The Concept of Energy,* 35).

8. Theobald, *The Concept of Energy,* 39.

9. Cited in Smith, *The Science of Energy,* 1.

10. Maxwell, "Hermann Ludwig Ferdinand Helmholtz," in *The Scientific Papers of James Clerk Maxwell* (1890), ed. W. D. Niven, 2 vols. (rpt. New York: Dover, n.d.), 2:594–95.

11. Stuart Peterfreund has expanded on this stark rhetorical background of the energy concept: "In the *Rhetoric,* Aristotle argues for the intrinsic power of figurative language to animate the inanimate and make present the absent through the properties of 'actuality' (Gr. *energeia*) and metaphor." "The Re-Emergence of Energy in the Discourse of Literature and Science," *Annals of Scholarship: Metastudies of the Humanities and Social Sciences* 4 (1986–87): 36.

12. On the relation of allegory to the discourse of the daemonic, see Clarke, *Allegories of Writing,* chap. 1 and passim.

13. See James T. Cushing, *Philosophical Concepts in Physics: The Historical Relation between Philosophy and Scientific Theories* (New York: Cambridge University Press, 1998).

14. Joel Fineman addresses the relation of doctrinal allegory to institutional power in "The Structure of Allegorical Desire," collected in *The Subjectivity Effect in Western Literary Tradition: Essays toward the Release of Shakespeare's Will* (Cambridge: MIT Press, 1991), 3–31.

15. Plato, *The Republic,* trans. Benjamin Jowett (New York: Modern Library, 1941), §514. The parable of the cave is part of a larger sequence in books 6 and 7 of the *Republic* figuring the relations between the phenomenal and the ideal worlds—the sun, the line, and the cave. This wider context underscores the cosmological and geometric subtexts embedded in the allegorical construction of the cave. See Julia Annas, "Understanding and the Good: Sun, Line, and Cave," in *Plato's "Republic": Critical Essays,* ed. Richard Kraut (New York: Rowman and Littlefield, 1997), 158; Andrew Barker and Martin Warner, eds., *The Language of the Cave* (Edmonton: Academic, 1992); and Seth Benardete, "Sun, Line, Cave," in *Socrates' Second Sailing: On Plato's "Republic"* (Chicago: University of Chicago Press, 1989), 157–77.

16. *The Republic,* §515.

17. In another context, Katherine Hayles terms this discursive maneuver the "Platonic backhand," which "works by inferring from the world's noisy multiplicity a simplified abstraction. So far so good: this is what theorizing should do. The problem comes when the move circles around to constitute the abstraction as the originary form from which the world's multiplicity derives." N. Katherine Hayles, *How We Became Posthuman: Virtual Bodies in Cybernetics, Literature, and Informatics* (Chicago: University of Chicago Press, 1999), 12.

18. *The Republic,* §517.

19. Luce Irigaray points out that in Plato's theater of phallic transcendentalism, "*Symmetry plays a decisive part* . . . as projection, reflection, inversion, retroversion—and you will always already have lost your bearings . . . though Socrates never

breathes a word about the whole mystification. This theatrical trick is unavoidable if you are to enter the functioning of representation." "Plato's Hystera," in *Speculum of the Other Woman,* trans. Gillian C. Gill (Ithaca, N.Y.: Cornell University Press, 1985), 244. See also Steven Shankman, ed., *Plato and Postmodernism* (Glenside, Pa.: Aldine Press, 1994); and Bruce Clarke, "History of Metamorphic Allegory," in *Allegories of Writing,* 23–51.

20. Cited in Smith, *The Science of Energy,* 168. See also Charles Coulston Gillispie, *The Edge of Objectivity: An Essay in the History of Scientific Ideas* (Princeton: Princeton University Press, 1960), 399; and P. M. Harman, *Energy, Force, and Matter: The Conceptual Development of Nineteenth-Century Physics* (New York: Cambridge University Press, 1982), 65.

21. For further discussion of such cultural factors, see Ted Underwood, "Productivism and the Vogue for 'Energy' in Late Eighteenth-Century Britain," *Studies in Romanticism* 34, no. 1 (1995): 103–25; and Maria Osietzki, "Rudolf Clausius: Entropy and Environment," in *Thermodynamics: History and Philosophy,* ed. K. Martinás, L. Ropolyi, and P. Szegedi (London: World Scientific, 1991), 57–62.

22. James Clerk Maxwell to P. G. Tait, February 12, 1872, in *Scientific Letters and Papers,* ed. Harman, 2:710 (Cambridge University Library, Add 7655/Ib/43).

23. James Clerk Maxwell to P. G. Tait, March 10, 1873, in *Scientific Letters and Papers,* ed. Harman, 2:834 (Cambridge University Library, Add 7655/Ib/52a). Maxwell picks up this idea in his review of Tait and Stewart's sequel to the *Unseen Universe, Paradoxical Philosophy:*

> There are methods . . . by which the application of energy may be directed without interfering with its amount. . . . The dynamical theory of a conservative material system shews us, however, that *in general* the present configuration and motion determine the whole course of the system, exceptions to this rule occurring only at the instants when the system passes through certain isolated and singular phases, at which a strictly infinitesimal force may determine the course of a system to any one of a finite number of equally possible paths, as the pointsman at a railway junction directs the train to one set of rails or another. (*Scientific Papers,* ed. Niven, 2:760)

See also P. M. Harman, *The Natural Philosophy of James Clerk Maxwell* (Cambridge: Cambridge University Press, 1998), 203–8.

24. Benjamin, *German Tragic Drama,* 165.

25. Balfour Stewart, *The Conservation of Energy, being an Elementary Treatise on Energy and its Laws,* 5th ed. (London: Kegan Paul, 1881), vii, 132.

26. See J. Edward Chamberlain, "Images of Degeneration: Turnings and Transformations," in *Degeneration: The Dark Side of Progress,* ed. J. Edward Chamberlain and Sander L. Gilman (New York: Columbia University Press, 1985), 263–89. See also Erwin N. Hiebert, "The Uses and Abuses of Thermodynamics in Religion," *Daedalus* 95, no. 4 (1966): 1046–80. For a lively survey of recent cosmological speculation, see Peter T. Landsberg, "From Entropy to God?" in *Thermodynamics: His-*

tory and Philosophy, ed. K. Martinás, L. Ropolyi, and P. Szegedi (London: World Scientific, 1991), 379–403.

27. Benjamin, *German Tragic Drama,* 178.

28. See Gay Clifford, *The Transformations of Allegory* (Boston: Routledge, 1974), 34. See also Fletcher, *Allegory,* 108–20.

29. Samuel Delany, "Is Cyberpunk a Good Thing or a Bad Thing?" *Mississippi Review* 47–48 (1988): 31, cited in Scott Bukatman, "Amidst These Fields of Data: Allegory, Rhetoric, and the Paraspace," *Critique: Studies in Contemporary Fiction* 33, no. 3 (1992): 200.

30. Brian McHale, *Postmodernist Fiction* (New York: Methuen, 1987), 45–46, cited in Bukatman, "Fields of Data," 207.

31. Rudolf Rucker, *Geometry, Relativity, and the Fourth Dimension* (New York: Dover, 1977), 4.

32. Charles Howard Hinton, *The Fourth Dimension* (1904; rpt. New York: Arno Press, 1976), 2–3.

33. See Stephen A. Barney, "Kinds of Allegory," in *Allegories of History, Allegories of Love* (Hamden, Conn.: Archon Books, 1978), 29–38.

34. "Transference by analogy" is "the situation that occurs whenever a second element is related to a first as a fourth is to a third. For the poet will then use the fourth in place of the second or the second in place of the fourth, and sometimes poets add the reference to which the transferred term applies. I mean, for example, that a cup is related to Dionysus as a shield is to Ares. The poet will, therefore, speak of the cup as the shield of Dionysus and the shield as the cup of Ares. . . . In some situations, there is no regular name to use to cover the analogous relation, but nevertheless the related elements will be spoken of by analogy; for example, to scatter seed is to sow, but the scattering of the sun's rays has no name. But the act of sowing in regard to grain bears an analogous relation to the sun's dispersing of its rays, and so we have the phrase 'sowing the god-created fire.'" Aristotle, *Poetics,* in *The Critical Tradition: Classic Texts and Contemporary Trends,* ed. David H. Richter (New York: St. Martin's, 1989), 58.

35. Cf. Luc Ferry's recent rehearsal of the analogical construction of the spatial fourth dimension in *Homo Aestheticus: The Invention of Taste in the Democratic Age,* trans. Robert De Loaiza (Chicago: University of Chicago Press, 1993), 212–14.

36. Edwin A. Abbott, *Flatland: A Romance of Many Dimensions* (1886; New York: Dover, 1992), 77.

37. G. F. Rodwell, "On Space of Four Dimensions," *Nature,* May 1, 1873, 9.

38. "And as she spoke his wet hair branched in antlers / Worn by the lively stag; his neck grew long, / Ears pointed, hands were hoofs, arms were thin legs, / And all his body a short-furred, spotted skin." Ovid, *The Metamorphoses,* trans. Horace Gregory (New York: Viking, 1958), book 3.

39. Abbott, *Flatland,* viii, 78.

40. Yevgeny Zamyatin, *We,* trans. Clarence Brown (New York: Penguin, 1993), 148.

41. See M. Keith Booker, *Dystopian Literature: A Theory and Research Guide*

(Westport, Conn.: Greenwood Press, 1994). Booker locates Plato's *Republic* and *We* in a rich international tradition of dystopian fictions (292–97). The cave itself in Plato's parable could be seen as the kernel of dystopia buried within the *Republic*'s ostensible utopia.

Chapter 2

1. "In the world of the symbol, it would be possible for the image to coincide with the substance. . . . Their relationship is one of simultaneity, which, in truth, is spatial in kind, and in which the intervention of time is merely a matter of contingency, whereas, in the world of allegory, time is the originary constitutive category." Paul de Man, "The Rhetoric of Temporality," in *Interpretation: Theory and Practice,* ed. Charles S. Singleton (Baltimore: Johns Hopkins University Press, 1969), 190.

2. Camille Flammarion, *La Fin du Monde* (Paris: Ernest Flammarion, 1894), translated as *Omega: The Last Days of the World* (New York: Cosmopolitan Publishing, 1894; rpt. New York: Arno Press, 1975).

3. Buck-Morss, *The Dialectics of Seeing,* 164. A complete translation has now appeared: Walter Benjamin, *The Arcades Project,* trans. Howard Eiland and Kevin Mclaughlin (Cambridge: Harvard University Press, 1999).

4. "Mythic time is, clearly, not limited to a particular discourse. Science as well as theology, rationalism as well as superstition can claim that events are inexorably determined. Nor are mythic explanations limited to a particular epoch. They have their (Western) source in classical antiquity and Biblical narrative. But they reappear in the most recent cosmological speculations" (Buck-Morss, *The Dialectics of Seeing,* 78). See also Greg Myers, "Nineteenth-Century Popularizations of Thermodynamics and the Rhetoric of Social Prophecy," in *Energy and Entropy: Science and Culture in Victorian Britain,* ed. Patrick Brantlinger (Bloomington: Indiana University Press, 1989), 307–38. Myers quotes Spengler: "What the myth of Gotterdammerung signified of old . . . the myth of entropy signifies today—*the world's end as completion of an inwardly necessary evolution,*" commenting: "The implication of this myth is that history is shaped by the laws of physics rather than by the struggles of people" (307).

5. On William Thomson's role in the theologizing of classical thermodynamics, see Crosbie Smith and M. Norton Wise, *Energy and Empire: A Biographical Study of Lord Kelvin* (New York: Cambridge University Press, 1989), esp. "The Irreversible Cosmos," 497–523; and Smith, *The Science of Energy,* esp. chap. 6.

6. Ilya Prigogine and Isabelle Stengers have commented on this mythic aspect of thermodynamic time in the nineteenth century: "the specific form in which time was introduced in physics, as a tendency toward homogeneity and death, reminds us more of ancient mythological and religious archetypes than of the progressive complexification and diversification described by biology and the social sciences" (*Order Out of Chaos,* 116). The mythic dimension of classical calculation, however, does not necessarily disqualify it as a functional mathematical idealization. The applica-

bility of reversible time depends on the nature and scale of the system under consideration. Prigogine and Stengers argue, "We must accept a pluralistic world in which reversible and irreversible processes coexist" (257). In current science, the mythic (repetitive, reversible) aspect of classical dynamics has now been repositioned in supplementary relation to the allegorical (irreversible) aspect of thermodynamics.

7. My purpose here is simply to elicit the allegorical character of Benjamin's "dialectical image," which remains a protean and contested concept that moves with equal facility in both verbal and visual registers. For a critique of visual bias in Buck-Morss's handling of the concept, see Chris Andre, "Aphrodite's Children: Hopeless Love, Historiography, and Benjamin's Dialectical Image," *SubStance* 85 (1998): 125–26.

8. Benjamin appropriated the mode of *phantasmagoria* from Marx's discussion of the commodity fetish. See Buck-Morss, *The Dialectics of Seeing,* 81.

9. Benjamin, *The Arcades Project,* 82.

10. Stewart, *Conservation of Energy,* 37, 47.

11. Balfour Stewart and P. G. Tait, *The Unseen Universe or Physical Speculations on a Future State* (New York: Macmillan, 1875), 81. See P. M. Heimann, "The *Unseen Universe:* Physics and the Philosophy of Nature in Victorian Britain," *British Journal for the History of Science* 21 (1972): 73–79.

12. Albert Einstein, "$E = MC^2$," in *Ideas and Opinions* (New York: Crown, 1954), 340.

13. Gregory Bateson, *Steps to an Ecology of Mind* (New York: Ballantine, 1972), xxi.

14. Karl Marx, "The Fetishism of Commodities," in *Karl Marx: Selected Writings,* ed. David McLellan (Oxford: Oxford University Press, 1977), 435. See also Christoph Asendorf, "Economy and Metamorphosis," in *Batteries of Life: On the History of Things and Their Perception in Modernity,* trans. Don Reneau (Berkeley and Los Angeles: University of California Press, 1993), 30–40; and Buck-Morss, *The Dialectics of Seeing,* 154.

15. Stewart, *Conservation of Energy,* 26–27; I am indebted here to Myers, "Nineteenth-Century Popularizations," esp. 326–29.

16. Asendorf, *Batteries of Life,* 154.

17. Stewart, *Conservation of Energy,* 97.

18. Stewart, *Conservation of Energy,* 144–45.

19. In "Thermodynamics and Sources of Solar Heat, 1846–1862," *British Journal of the History of Science* 15 (1982): 155–81, Frank James notes that in the 1850s, the scientific "realisation that the sun was the ultimate source of all energy for man[,] while not an original sentiment, seems in the thought of the founders of thermodynamics to have taken on a special significance. As soon as it was realised by them that . . . the sun was running down, it became important to them to discover the source of its energy" (174). See also Stephen G. Brush, *The Temperature of History: Phases of Science and Culture in the Nineteenth Century* (New York: Burt Franklin, 1978).

20. Unpublished letter to Richard Monckton Milnes, November 19, 1862, cited

in Dinah Birch, "'The Sun Is God': Ruskin's Solar Mythology," in *The Sun Is God: Painting, Literature, and Mythology in the Nineteenth Century,* ed. J. B. Bullen (Oxford: Clarendon Press, 1989), 112.

21. William Thomson, "On the Age of the Sun's Heat," *Macmillan's,* March 1862, 393. See Gillian Beer, "'The Death of the Sun': Victorian Solar Physics and Solar Myth," in Bullen, *The Sun Is God,* 159–80. That this idea persisted well into the twentieth century is attested to by Jacquetta Hawkes. Writing in 1962 about her upbringing in the decade of the Great War, she related that "when I was a child it was thought that as the sun was pouring out its treasure of heat, light and other forms of energy with such royal munificence it must gradually exhaust its resources and cool down. I remember feeling real grief as I pictured earth losing the warmth of its star and becoming more and more arctic until at last all life had withered away and it spun in darkening space, a grey and frozen planet." *Man and the Sun* (London: Cresset Press, 1963), 18.

22. All passages cited are from Edward Bulwer-Lytton, *The Coming Race, Falkland, Zicci,* and *Pausanias the Spartan* (London: Routledge, n.d.). See *The Definitive Time Machine,* ed. Harry M. Geduld (Bloomington: Indiana University Press, 1987), 3. See also Brian W. Aldiss, *The Billion Year Spree* (New York: Doubleday, 1973), 83–84; and James L. Campbell, "Edward Bulwer-Lytton's *The Coming Race* as a Condemnation of Advanced Ideas," *Essays in Arts and Sciences* 16 (May 1987): 55–63.

23. Bulwer-Lytton, *The Coming Race,* 12.

24. Bulwer-Lytton, *The Coming Race,* 45.

25. Müller's considerable work on ancient languages and mythologies aspired to scientific status but is recognizable today as a form of positivist Victorian scientism: see Beer, "Death of the Sun," 164.

26. See Birch, "'The Sun Is God,'" 118–20.

27. Bulwer-Lytton, *The Coming Race,* 14.

28. The narrator mentions the electrical researches of Faraday and Maxwell, as well as several electromagnetic scientisms prominent at midcentury—mesmerism, electrobiology, and odic force.

29. Bulwer-Lytton, *The Coming Race,* 88.

30. Cf. Bruce Mazlish, "A Triptych: Freud's *The Interpretation of Dreams,* Rider Haggard's *She,* and Bulwer-Lytton's *The Coming Race,*" *Comparative Studies in Society and History* 35, no. 4 (1993): 726–45.

31. Bulwer-Lytton, *The Coming Race,* 66.

32. Bulwer-Lytton, *The Coming Race,* 68.

33. Bulwer-Lytton, *The Coming Race,* 48.

34. Flammarion, *Omega,* 167.

35. Flammarion, *Omega,* 9.

36. Flammarion, *Omega,* 43.

37. Flammarion, *Omega* 43–44. With regard to the financial speculator's "rolling" out of the institute's upper gallery, Asendorf notes the "rolling" of capital: some cultures have manufactured currency in the form of little balls, their spherical

form reinforcing their ability to "circulate" in all directions, or globally (*Batteries of Life,* 11–12). The racial detail also reinforces the note of the international circulation of capital in the finance of commercial trade.

38. Flammarion, *Omega,* 178.

39. Flammarion, *Omega,* 189; emphasis added.

Chapter 3

1. Latour, *Science in Action,* 141.

2. Iain Cameron and David O. Edge, *Scientific Images and Their Social Uses: An Introduction to the Concept of Scientism* (Boston: Butterworth, 1979), 3. For a historical evaluation of the concept, see Casper Hakfoort, "The Historiography of Scientism: A Critical Review," *History of Science* 33 (1995): 375–95.

3. Rabinbach, *The Human Motor,* 10.

4. Donna J. Haraway, *Modest_Witness@Second_Millennium.FemaleMan© _Meets_OncoMouse™: Feminism and Technoscience* (New York: Routledge, 1997), 114.

5. For "the early pioneers of behavioral science," Bateson wrote, "it was tempting to relate 'energy' to already existing metaphors such as 'strength' of emotions or character or 'vigor.' Or to think of 'energy' as somehow the opposite of 'fatigue' or apathy.' Metabolism obeys an energy budget (within the strict meaning of 'energy'), and energy expended in behavior must surely be included in this budget; therefore it seemed sensible to think of energy as a determinant of behavior" (*Steps to an Ecology,* xxi–xxii).

6. Ferry, *Homo Aestheticus,* 215.

7. Karl Marx, preface to *Capital,* vol. 1, in *Selected Writings,* 416.

8. Marx, preface to *Capital,* 417.

9. Bruno Latour, *Aramis; or, The Love of Technology,* trans. Catherine Porter (Cambridge: Harvard University Press, 1996), 119.

10. Bruno Latour, *We Have Never Been Modern,* trans. Catherine Porter (Cambridge: Harvard University Press, 1993), 96, 95.

11. "Commonplaces are useful rhetorically both because while they may have a great deal of evidence to support them in specific contexts, they need not be supported, and because while they may have a quite definite meaning in one discourse, they transfer easily to other discourses" (Myers, "Nineteenth-Century Popularizations," 309).

12. Haraway, *Modest Witness,* 43–44.

13. Martin Jay, "Modernism and the Specter of Psychologism," *Modernism/Modernity* 3, no. 2 (1996): 93.

14. Pierre Bourdieu, *The Field of Cultural Production: Essays on Art and Literature,* ed. and intro. Randal Johnson (New York: Columbia University Press, 1993), 30.

15. Hayles, *Chaos Bound,* xi.

16. Hayles, *Chaos Bound,* 33.

17. Hayles, *Chaos Bound,* 39.

18. On the midcentury impress of natural theology, see Silvan S. Schweber, "Demons, Angels, and Probability: Some Aspects of British Science in the Nineteenth Century," in *Physics as Natural Philosophy,* ed. Abner Shimony and Herman Feshbach (Cambridge: MIT Press, 1982), 332–33. See also Donald E. Hall, ed., *Muscular Christianity: Embodying the Victorian Age* (Cambridge: Cambridge University Press, 1995).

19. Thomson, "Universal Tendency," 1:514.

20. "Maxwell and his allies placed the doctrine of dissipation, embodied solidly in scientific and engineering practice, center stage in the science of energy. It was dissipation of energy which gave fresh meaning to the Christian perspective on beginnings and endings to the visible cosmos. By pointing to the necessity of a beginning, the original divine creation of matter and of energy, the doctrine purported to undermine all those creeds which espoused a self-regulating, self-sustaining and eternal material world" (Smith, *The Science of Energy,* 239–40).

21. On the problem of "continuity" in Victorian science and culture, see Donald R. Benson, "'Catching Light': Physics and Art in Walter Pater's Cultural Context," in *One Culture: Essays in Science and Literature,* ed. George Levine and Alan Rauch (Madison: University of Wisconsin Press, 1987), 143–63, and "Facts and Fictions in Scientific Discourse: The Case of Ether," *Georgia Review* 38 (1984): 825–37.

22. Smith, *The Science of Energy,* 240.

23. Cited in Myers, "Nineteenth-Century Popularizations," 324.

24. Smith, *The Science of Energy,* 6 and passim.

25. Herbert Spencer to John Tyndall, no date (c. 1858). Courtesy of the Royal Institution of Great Britain, Tyndall Correspondence, 3:1175 (19/D1.90).

26. Smith, *The Science of Energy,* 111.

27. Rabinbach, *The Human Motor,* 9.

28. Rabinbach, *The Human Motor,* 4.

29. In *The Human Motor,* Rabinbach discusses the interplay of thermodynamics with concurrent mechanistic approaches to the "human motor," the body itself as a producer, consumer, and converter of energies.

30. Gillispie, *The Edge of Objectivity,* 405.

31. Gillispie, *The Edge of Objectivity,* 405.

32. Henry Adams, *A Letter to American Teachers of History* (1910), in *The Degradation of the Democratic Dogma* (New York: Macmillan, 1920), 196–97.

33. Ostwald's full-blown cultural system was based precisely on a visionary moment in 1890, sitting in the transfusing sunshine of the Berlin Tiergarten the morning after a long evening of heated collegial debate about energy: "Nobody, probably not even Ostwald himself, could have suspected that twenty years later, shortly after receiving the Nobel Prize, he would begin writing more than a hundred weekly 'Sonntagspredigten' (Sunday sermons) in which he disseminated not only a popular version of energetics, but also the energeticist world-view he had developed in the intervening years." Casper Hakfoort, "Science Deified: Wilhelm Ostwald's

Energeticist World-View and the History of Scientism," *Annals of Science* 49 (1992): 527.

34. Adams, *Letter to American Teachers,* 163.

35. Cited in Adams, *Letter to American Teachers,* 168–69.

36. As a scientific theory, vitalism prospered in the disciplines of chemistry and medicine, which "were, in the late eighteenth century, privileged sciences for those who fought against the physicists' *esprit de système* in favor of a science that would take into account the diversity of natural processes" (Prigogine and Stengers, *Order Out of Chaos,* 83). See the section "Vitalism, Evolution, and Thermodynamics" in my *Dora Marsden,* 26–32.

37. Edward Carpenter, *Civilisation: Its Cause and Cure and Other Essays,* 12th ed. (London: George Allen, 1912), 36. For the larger context of Carpenter's remarks, see "Edward Carpenter's Evolutionary Sunshine," in my *Dora Marsden,* 32–38.

38. Adams, *Letter to American Teachers,* 145. "Ruskin sees that Tyndall is using the language of Carlyle and Homer for a materialist argument. Adams, on the other hand, sees Tyndall as shoring up traditional ideas. This 'famous outburst of eloquence' [the peroration of Tyndall's *Heat as a Mode of Motion*] is the last gasp of 'the scientific dogma which asserted resolutely, without qualification, the fact that nothing in nature was lost' (Adams, *Letter,* p. 143)" (Myers, "Nineteenth-Century Popularizations," 316). On the North British dissatisfaction with the metropolitan Tyndall's minimizing of the second law, see Smith, *The Science of Energy,* 256.

39. Adams, *Letter to American Teachers,* 146.

40. Dora Marsden, "The Over-and-Above in Life," *Freewoman,* February 1, 1912, 201. See my *Dora Marsden,* 52.

41. Marsden, "Over-and-Above," 201–2.

42. Marsden, "Over-and-Above," 202.

43. D. H. Lawrence, *The Rainbow* (New York: Penguin, 1976), 436.

44. Lawrence, *The Rainbow,* 440–41.

45. The late-nineteenth-century American philosopher of science J. B. Stallo characterized the physical sources of Dr. Frankstone's scientific dogma in this way: "the ultimate problem of the atomo-mechanical theory, to whose effectual and complete solution the physicists of the day look forward with a greater or less degree of confidence . . . is the exhibition of all vital and organic phenomena as results of ordinary chemical and physical action, and of chemical and physical action, in turn, as exchanges and transferences of mechanical motion between constant and uniform elements of mass." J. B. Stallo, *The Concepts and Theories of Modern Physics* (London: Kegan Paul, Trench, 1882), 270.

46. Lawrence, *The Rainbow,* 441.

47. Ibid.

48. See Smith, *The Science of Energy,* 151.

49. Yevgeny Zamyatin, "On Literature, Revolution, Entropy, and Other Matters," in *A Soviet Heretic: Essays by Yevgeny Zamyatin,* ed. and trans. Mirra Ginsburg (Chicago: University of Chicago Press, 1970), 107–8. This major essay is sub-

mitted to a rather garbled deconstruction in Michael Beehler, "Yevgeny Zamyatin: The Essential, the Superfluous, and Textual Noise," *SubStance* 15:2 (1986): 48–60.

50. Zamyatin, "Literature, Revolution, Entropy," 112.

51. Zamyatin, "Literature, Revolution, Entropy," 108.

52. Zamyatin, "Literature, Revolution, Entropy," 107.

53. Casper Hakfoort gives convincing evidence that positivist scientistic productions operate with "the authority of science [to fulfill] psychological functions similar to the ones served by religion" ("The Historiography of Scientism," 387).

Chapter 4

1. See my *Allegories of Writing,* 1–13.

2. On the informatic forms of contemporary messenger figures, see Michel Serres, *Angels: A Modern Myth* (New York: Flammarion, 1995); and Davis, *Techgnosis.*

3. Edwardian author Allen Upward evoked the scientific daemonic while satirizing what he saw as the "Latinate" pretensions of modern scientific terminology in the following spoof on the concept of potential energy—in particular, the potential energies conferred upon material objects strictly by their relative positions within a (gravitational) field. According to the "priests" of science, "when you throw the stone down there is some other power at work besides your push and the earth's pull; there is this mysterious Potential Energy which has been inside the stone all the time. It seems to be a scientific case of demon-possession. The demon of Latin [i.e., potential, from Latin *potentia,* power] Energy enters into some stones but not others. It prefers stones on the roof of a house if it can get them, but if not, it will take stones on mountains." *The New Word* (New York: Mitchell Kennerly, 1910), 134. See also Jerome Rothstein, "Physical Demonology," *Methodos* 42 (1959): 99–121.

4. On Laplace's Demon, see Ernst Cassirer, *Determinism and Indeterminism in Modern Physics* (New Haven: Yale University Press, 1956), 3–25; and Prigogine and Stengers, *Order Out of Chaos,* 75–77.

5. For more on Victorian responses to thermodynamics, see Peter Allan Dale, *In Pursuit of a Scientific Culture: Science, Art, and Society in the Victorian Age* (Madison: University of Wisconsin Press, 1989), 130–36, 225–32; George Levine, "*Little Dorrit* and Three Kinds of Science," in *Dickens and Other Victorians,* ed. Joanne Shattock (New York: St. Martin's, 1988), 3–24; Myers, "Nineteenth-Century Popularizations"; and Beer, "Wave Theory."

6. James Clerk Maxwell, *Theory of Heat,* 9th ed. (London: Longmans, Green, 1888), 328–29.

7. Fletcher, *Allegory,* 39.

8. On the perennial relation of thermodynamics to ideas of cultural crisis, see L. Ropolyi, "Thermodynamic Elements in World Views—World View Elements in Thermodynamics," in Martinás, Ropolyi, and Szegedi, *Thermodynamics,* 424–39.

9. Harvey S. Leff and Andrew F. Rex, eds., *Maxwell's Demon: Entropy, Information, Computing* (Princeton: Princeton University Press, 1990). See also Werner

Ehrenberg, "Maxwell's Demon," *Scientific American* 217 (1967): 103–10; and John D. Collier, "Two Faces of Maxwell's Demon Reveal the Nature of Irreversibility," *Studies in the History and Philosophy of Science* 21, no. 2 (1990): 257–68.

10. Fletcher, *Allegory,* 39.

11. See Deborah L. Madsen, *The Postmodernist Allegories of Thomas Pynchon* (New York: St. Martin's, 1991): "Paranoia is the neurosis most akin to a figural hermeneutic" (23).

12. Thomas Pynchon, *The Crying of Lot 49* (New York: Harper and Row, 1990), 24–25.

13. For a contemporary example of the cultural allegorization of mathematical concepts, see Robert M. Markley's discussion of cyberspace in the context of boundary mathematics, in "Boundaries: Mathematics, Alienation, and the Metaphysics of Cyberspace," *Configurations* 3:3 (Fall 1994): 485–507.

14. Cf. Donna Haraway on the "leaky distinction" between organism and machine: "Pre-cybernetic machines could be haunted; there was always the spectre of the ghost in the machine. This dualism structured the dialogue between materialism and idealism that was settled by a dialectical progeny, called spirit or history, according to taste." "A Cyborg Manifesto," in *Simians, Cyborgs, and Women: The Reinvention of Nature* (New York: Routledge, 1991), 152. Postcybernetic machines—computers—are no longer haunted; they are themselves the ghosts and demons they house, reified in silicon.

15. Pynchon, *The Crying of Lot 49* 85–86. See N. Katherine Hayles, "'A Metaphor of God Knew How Many Parts': The Engine That Drives *The Crying of Lot 49,*" in *New Essays on "The Crying of Lot 49,"* ed. Patrick O'Donnell (New York: Cambridge University Press, 1991), 97–125; Charles H. Bennett, "Demons, Engines, and the Second Law," *Scientific American* 257 (1987): 108–16; and Peter L. Abernethy, "Entropy in Pynchon's *The Crying of Lot 49,*" *Critique* 14, no. 2 (1974): 18–32.

16. See William R. Paulson, *The Noise of Culture: Literary Texts in a World of Information* (Ithaca, N.Y.: Cornell University Press, 1988), 54–58; and Hayles, *Chaos Bound,* 48–54, and "'Metaphor of God,'" 109–13.

17. Pynchon, *Crying of Lot 49,* 106.

18. Gillian Beer describes the subtlety of Maxwell's position: "Maxwell combines the extreme of skepticism with the extreme of faith, remaining always devout while teasing out the obduracies of the invisible material world. . . . He was highly conscious of the changing functions of metaphor as they extend across scientific fields" ("Wave Theory," 205).

19. See the "Catechism Concerning Demons," in Cargill Gilston Knott, *Life and Scientific Work of Peter Guthrie Tait* (Cambridge: Cambridge University Press, 1911), 214–15. James Clerk Maxwell to John William Strutt, December 6, 1870, in *Scientific Letters and Papers,* ed. Harman, 2:583. See also "The Demon," in Smith and Wise, *Energy and Empire,* 621–33.

20. See David Wilson, "The Educational Matrix: Physics Education at Early-Victorian Cambridge, Edinburgh, and Glasgow Universities," in *Wranglers and Physi-*

cists: Studies on Cambridge Physics in the Nineteenth Century, ed. P. M. Harman (Manchester: Manchester University Press, 1985), 12–48; P. M. Harman, "Edinburgh Philosophy and Cambridge Physics: The Natural Philosophy of James Clerk Maxwell," in *Wranglers and Physicists,* 202–24; and Richard Olson, *Scottish Physics and British Physicists* (Princeton: Princeton University Press, 1975), 287–321.

21. C. W. F. Everitt, "Maxwell's Scientific Creativity," in *Springs of Scientific Creativity: Essays on the Founders of Modern Science,* ed. Rutherford Aris, H. Ted Davis, and Roger H. Stuewer (Minneapolis: University of Minnesota Press, 1983), 93.

22. Cited in Knott, *Life of Tait,* 175.

23. "A Vision of a Wrangler, of a University, of Pedantry, and of Philosophy," in Campbell and Garnett, *Life of Maxwell,* 612–17.

24. Martin Goldman, *The Demon in the Aether: The Story of James Clerk Maxwell* (Edinburgh: Paul Harris, 1983), 17.

25. Maxwell, *Scientific Letters and Papers,* ed. Harman, 1:319–20.

26. On the development of "Maxwell's equations" for the electromagnetic field, see Bruce J. Hunt, *The Maxwellians* (Ithaca, N.Y.: Cornell University Press, 1991).

27. Letter from Thomson to Faraday, June 11, 1847, in S. P. Thompson, *The Life of William Thomson, Baron Kelvin of Largs,* 2 vols. (London, 1910), 1:203.

28. "Address to the Mathematical and Physical Sections of the British Association" (1870), in *Scientific Papers,* ed. Niven, 2:218.

29. James Paxson discusses the inherent allegorical quality of mathematical discourse in "The Allegory of Temporality and the Early Modern Calculus," *Configurations* 4, no. 1 (1996): 42.

30. Thomson's cosmological speculations stemmed originally from his work on the theory of solar heat. See James, "Thermodynamics and Sources," 161–63; and David B. Wilson, "Kelvin's Scientific Realism: The Theological Context," *Philosophical Journal* 11 (1974): 41–60.

31. Smith and Wise, *Energy and Empire,* 625. Pynchon's Nefastis retains this doctrine.

32. James Clerk Maxwell, "On Faraday's Lines of Force," in *Scientific Papers,* ed. Niven, 1:156.

33. Maxwell, "Faraday's Lines of Force," 1:156–57.

34. William Berkson, *Fields of Force: The Development of a World View from Faraday to Einstein* (New York: John Wiley and Sons, 1974), 137. For more discussion and documentation of scholarship on Maxwell's analogical methods, see Daniel M. Siegel, "Mechanical Image and Reality in Maxwell's Electromagnetic Theory," in Harman, *Wranglers and Physicists,* 180–201.

35. See Margaret Morrison, "A Study in Theory Unification: The Case of Maxwell's Electromagnetic Theory," *Studies in the History and Philosophy of Science* 23, no. 1 (1992): 107.

36. Berkson, *Fields of Force,* 144.

37. Maxwell, *Scientific Papers,* ed. Niven, 1:187.

38. Siegel, "Mechanical Image," 191.

39. On the Apostles, Everitt comments, "The Apostles were primarily classicists rather than mathematicians. Maxwell was one of the rare exceptions, although years later two other Cambridge mathematicians of note, A. N. Whitehead and Bertrand Russell, also entered the charmed circle" ("Maxwell's Scientific Creativity," 103). See also Robert O. Preyer, "The Romantic Tide Reaches Trinity: Notes on the Transmission and Diffusion of New Approaches to Traditional Studies at Cambridge, 1820–1840," in *Victorian Science and Victorian Values: Literary Perspectives,* ed. James Paradis and Thomas Postlewait (New York: New York Academy of Sciences, 1981), 39–68.

40. Maxwell, "Are There Real Analogies in Nature?" in Campbell and Garnett, *Life of Maxwell,* 237–38.

41. On Kant's concept of space in the context of the fourth dimension, see Henderson, *Fourth Dimension,* 10–18. For Maxwell on the fourth dimension, see Alfred M. Bork, "The Fourth Dimension in Nineteenth-Century Physics," *Isis* 55, no. 3 (1964): 328–30.

42. Campbell and Garnett, *Life of Maxwell,* 239.

43. Campbell and Garnett, *Life of Maxwell,* 236.

44. "Ether carried most of the marks of a scientific fact: it performed what appeared to be a necessary and fundamental role in the given and developing system of nature; it was described and measured; it was widely taken for fact. Yet it bore even more basic marks of a fiction: it was an imaginative construct, derived in significant part by metaphor and analogy, and taken for a fiction by some scientists (including certain of those who, paradoxically, also took it for a fact)" (Benson, "Facts and Fictions," 836–37).

45. See Robert Kargon, "Model and Analogy in Victorian Science: Maxwell's Critique of the French Physicists," *Journal of the History of Ideas* 30 (1969): 433; Harman, *Energy, Force, and Matter,* 79–80; and Siegel, "Mechanical Image," 192–93.

46. Berkson, *Fields of Force,* 147, 152.

47. Siegel, "Mechanical Image," 188–94.

48. Berkson, *Fields of Force,* 149.

49. "This identification of the magneto-electric and luminiferous media was not yet an electromagnetic theory of light, for the 'transverse undulations' constituting light waves were not given any definite interpretation in terms of electromagnetic variables; what Maxwell had accomplished was not a reduction of optics to electricity and magnetism but rather a reduction of both to the mechanics of one ether." Daniel M. Siegel, "Thomson, Maxwell, and the Universal Ether in Victorian Physics," in *Conceptions of Ether: Studies in the History of Ether Theories 1740–1900,* ed. G. N. Cantor and M. J. S. Hodge (New York: Cambridge University Press, 1981), 254. See also Nancy J. Nersessian, "Kuhn and the Cognitive Revolution," in *Configurations* 6, no. 1 (1998): 112–14.

50. Barney, "Kinds of Allegory," 33.

51. Maxwell, *Scientific Papers,* ed. Niven, 1:500.

52. Goldman, *Demon in the Aether,* 155.

53. Kargon, "Model and Analogy," 434.

54. For the background of this volume, see Heimann, "The *Unseen Universe.*"

55. Tait and Stewart, *Unseen Universe,* 63–64.

56. P. G. Tait, *Sketch of Thermodynamics* (Edinburgh: Edmonston and Douglas, 1868).

57. On the prehistory of Tait's *Sketch of Thermodynamics,* see Knott, *Life of Tait,* 208–14.

58. P. G. Tait to James Clerk Maxwell, December 6, 1867, in *Maxwell on Heat and Statistical Mechanics: On "Avoiding All Personal Enquiries" of Molecules,* ed. Elizabeth Garber, Stephen G. Brush, and C. W. F. Everitt (Bethlehem, Pa.: Lehigh University Press, 1995), 175 (Cambridge University Library, Add 7655/Ia/4).

59. James Clerk Maxwell to P. G. Tait, December 11, 1867, in *Maxwell on Heat,* 176. See also Smith and Wise, *Energy and Empire,* 621–22.

60. Maxwell to Tait, December 11, 1867, 177.

61. Maxwell to Tait, December 11, 1867, 177.

62. Maxwell to Tait, December 11, 1867, 177.

63. "The Demon is a serious intellectual construct that has played a nonnegligible conceptual role in the natural sciences. But is it not also a theorization of nineteenth-century fantasies—perpetual motion, energy without expenditure, the production of a generalized reservoir?" (Paulson, *The Noise of Culture,* 41–42).

64. Maxwell to Tait, December 11, 1867, 177.

65. Concerning this annotation, identified as Thomson's by Knott (*Life of Tait,* 214), the editors of *Maxwell on Heat* state that they "consider the writing closer to Tait's than to Thomson's" (178). But Tait placed his own annotation on the manuscript of Maxwell's letter—a note to Thomson, to which the remarks in question appear to be a reply. My examination of the original, as well as the congruence of the annotation with Thomson's 1874 paper, would support Knott's prior identification.

66. William Thomson, "The Kinetic Theory of the Dissipation of Energy" (1874) and "The Sorting Demon of Maxwell" (1879), in *Mathematical and Physical Papers,* ed. Sir Joseph Larmor (Cambridge: Cambridge University Press, 1911), 5:11–23.

67. Thomson, "Kinetic Theory," 11.

68. Thomson, "Kinetic Theory," 12.

69. Thomson's remarks promoting his nomination make it clear that he had the Platonic or mediating form of the daemonic (and not the Judeo-Christian demonic) in mind. Putting his own words in Maxwell's mouth, Thomson placed this footnote to his phrase "Maxwell's 'intelligent demons' ": "The definition of a demon, according to the use of this word by Maxwell, is an intelligent being endowed with free-will and fine enough tactile and perceptive organization to give him the faculty of observing and influencing individual molecules of matter" ("Kinetic Theory," 12). Thomson demonstrated familiarity with the classical hermeneutics of allegory in "The Sorting Demon of Maxwell": "The word 'demon,' which originally in Greek meant

a supernatural being, has never been properly used to signify a real or ideal personification of malignity" (21).

70. Tait and Stewart, *Unseen Universe,* 86.

71. Clearly Tait and Stewart were responding to Thomson's 1874 article "The Kinetic Theory of the Dissipation of Energy." In this spirited treatment, Thomson recruited "an army of Maxwell's 'intelligent demons,'" and so initiated a line of martial metaphor alien to Maxwell's own discourse. Thomson went on: "Now, suppose the weapon of the ideal army to be a club, or, as it were, a molecular cricket bat; and suppose, for convenience, the mass of each demon with his weapon to be several times greater than that of a molecule. . . . The whole interface between hot and cold is to be divided into small areas, each allotted to a single demon. The duty of each demon is to guard his allotment, turning molecules back, or allowing them to pass through from either side, according to certain definite orders" (12–13). This imperialistic or teamwork regimentation of the demon-fiction is a striking example of the facility with which Maxwell's creation lends itself to further allegorization.

72. Tait and Stewart, *Unseen Universe,* 89.

73. Tait and Stewart, *Unseen Universe,* 90.

74. Tait and Stewart, *Unseen Universe,* 64.

Chapter 5

1. For the wider social context of Hinton's writings, see Paul Fayter, "Strange New Worlds of Space and Time: Late Victorian Science and Science Fiction," in *Victorian Science in Context,* ed. Bernard Lightman (Chicago: University of Chicago press, 1997), 256–80.

2. Charles Howard Hinton, "What is the Fourth Dimension?" in *Scientific Romances,* 1st series (London: Swan Sonnenschein, 1886; rpt. with 2d series, New York: Arno Press, 1976), 1:3–32. See Henderson, *Fourth Dimension,* 24–25; Rucker, *Geometry, Relativity,* 4; and James Webb, introduction to Hinton, *Scientific Romances,* n.p.

3. Abbott, *Flatland,* 62.

4. Elliot L. Gilbert, "'Upward, Not Northward': *Flatland* and the Quest for the New," *English Literature in Transition* 34, no. 4 (1991): 391–404; and Jonathan Smith, "'Euclid Honourably Shelved': Edwin Abbott's *Flatland* and the Methods of Non-Euclidean Geometry," in *Fact and Feeling: Baconian Science and the Nineteenth-Century Literary Imagination* (Madison: University of Wisconsin Press, 1994), 180–210.

5. Hinton, "A Plane World," in *Scientific Romances,* 1:129.

6. On James Hinton, the author of prose works promoting universal altruism popular with social radicals such as Edward Carpenter and the Havelock Ellises, see Marvin H. Ballard, "The Life and Thought of Charles Howard Hinton," M.A. thesis, Virginia Polytechnic Institute, 1980, 6–12; and Edith Ellis, *Three Modern Seers* (New York: Mitchell Kennerly, 1910). "The father James Hinton was an ear surgeon who was best known for *The Mystery of Pain,* a little book which sets forth the

Panglossian thesis that 'all that which we feel as painful is really *giving*—something that our fellows are better for, even though we cannot trace it.' It gives some idea of the turn of the son Charles Hinton's mind to learn that he wrote a piece, 'The Persian King,' in which he attempted to use higher dimensions and infinite series to obtain a mathematically accurate model of this idea." Rudy Rucker, ed., *Speculations on the Fourth Dimension: Selected Writings of Charles H. Hinton* (New York: Dover, 1980), v.

7. Hinton, "What is the Fourth Dimension?" 24.

8. Hinton, "The Recognition of the Fourth Dimension," *Bulletin of the Philosophical Society of Washington* 14 (1902); rpt. in *Speculations on Fourth Dimension,* ed. Rucker, 142–62. Hinton includes this essay in *The Fourth Dimension* (1904) under the title "A Recapitulation and Extension of the Physical Argument."

9. Hinton, *The Persian King,* in *Scientific Romances,* 1:38–39.

10. Hinton, *The Persian King,* 1:42.

11. Hinton, *The Persian King,* 1:105.

12. Hinton, *The Persian King,* 1:44.

13. Hinton, *The Persian King,* 1:50–51.

14. See Fletcher's "Daemonic mechanism and allegorical 'machines'": "Constriction of meaning, when it is the limit put upon a personified force or power, causes that personification to act somewhat mechanistically. The perfect allegorical agent is not a man possessed by a daemon, but a robot" (*Allegory,* 55).

15. Hinton, *The Persian King,* 1:59.

16. Hinton, *The Persian King,* 1:69. "Eighteenth-century physicists had considered energy losses within mechanical systems to be isolated from nonmechanical processes, and therefore they did not enunciate a theory of the equivalence of heat and mechanical energy. The concept of the conservation of mechanical energy was, however, to be found in eighteenth-century treatises on mechanics; it was familiar, in the form first enunciated by Leibniz, as the principle of the conservation of *vis viva* ('living force')" (Harman, *Energy, Force, and Matter,* 35–36).

17. Thomson "Universal Tendency," 514.

18. Hinton, *The Persian King,* 1:77.

19. Hinton, *The Persian King,* 1:80.

20. See Rabinbach, *The Human Motor;* and Hakfoort's discussion of Wilhelm Ostwald, "Science Deified."

21. Hinton, *The Persian King,* 1:99.

22. Hinton, *The Persian King,* 1:108.

23. See my "Resistance in Theory and the Physics of the Text," *New Orleans Review* 18, no. 1 (1991): 86–93.

24. Hinton, *The Persian King,* 1:107–8.

25. Hinton, *The Persian King,* 1:121.

26. Henderson, *Fourth Dimension,* 28; Hinton, *A New Era of Thought* (1888; London: Swan Sonnenschein, 1910), 1–4, and *The Fourth Dimension* (1904), 107–21.

27. Hinton, *The Persian King,* 1:122.

28. See Michio Kaku, *Hyperspace: A Scientific Odyssey through Parallel Universes, Time Warps, and the Tenth Dimension* (New York: Oxford University Press, 1994), 55–71; Stuart Moulthrop, "Paragnosis, or the Story of Cyberspace," *Works and Days* 12, nos. 1–2 (1994): 33–55; and Tony Robbin, *Fourfield: Computers, Art, and the Fourth Dimension* (Boston: Little, Brown, 1992).

29. H. G. Wells, *The Time Machine* (New York: Random House, 1931), ix–x.

30. Latour, *Aramis,* 222.

31. H. G. Wells, *The Time Machine,* ed. Michael Moorcock (Rutland, Vt.: Charles E. Tuttle, 1993), 3.

32. See Maureen Quilligan, *The Language of Allegory: Defining the Genre* (Ithaca, N.Y.: Cornell University Press, 1979), 22.

33. Quilligan cites Hester Prynne's emergence at the doorway of the courthouse in *The Scarlet Letter:* "More than an emblem for sexuality, the rosebush [beside the door] becomes a signal for the reader to begin looking for a moral, and to begin reading allegorically" (*The Language of Allegory,* 52). Cf. Oedipa Maas's first vista of San Narciso in Pynchon's *The Crying of Lot 49.*

34. Wells, *The Time Machine,* ed. Moorcock, 21. See Frank Scafella, "The White Sphinx and *The Time Machine,*" *Science Fiction Studies* 8, no. 3 (1981): 255–65, and "From Reason to Intelligence: Wells's White Sphinx as Chronotope of Nineteenth Century Science," *Cahiers Victoriens et Edouardiens* 46 (1997): 181–89.

35. Serres, *Hermes,* 48, 47.

36. Serres, *Hermes,* 43.

37. See Kathryn Hume, "Eat or Be Eaten: H. G. Wells's *Time Machine,*" *Philological Quarterly* 69, no. 2 (1990): 233–51, for an incisive reading of the way that ideological pressures render the novella's social message incoherent.

38. Wells's extrapolations of problematic biological and physical theories in this and other early tales lay down a generic pattern for the construction of science fictions through the operation of critical scientisms within literary narratives. See Darko Suvin, "A Grammar of Form and a Criticism of Fact: *The Time Machine* as a Structural Model for Science Fiction," in *H. G. Wells and Modern Science Fiction,* ed. Darko Suvin and Robert M. Philmus (Lewisburg, Pa.: Bucknell University Press, 1976), 90–115; and Darko Suvin, *Metamorphoses of Science Fiction: On the Poetics and History of a Literary Genre* (New Haven: Yale University Press, 1979), 208–42.

39. Simon Newcomb, "Modern Mathematical Thought," *Nature,* February 1, 1894, 325–29: "Now it is a fundamental principle of pure science that the liberty of making hypotheses is unlimited. . . . Add a fourth dimension to space, and there is room for an indefinite number of universes" (328).

40. Wells, *The Time Machine,* ed. Moorcock, 4–5.

41. In "Deconstructing the Time Machine," *Science-Fiction Studies* 14, no. 2 (1987): 201–21, Veronica Hollinger borrows Roland Barthes's distinction between the "Newtonian" time of the Work and the "Einsteinian" time of the Text to argue, "In many ways, *The Time Machine* appears to be an exemplary demonstration of Time as Newtonian Work" (207).

42. Wells, *The Time Machine,* ed. Moorcock, 19.

43. Cf. Robert M. Philmus, "The *Time Machine;* or, the Fourth Dimension as Prophecy," *PMLA* 84 (1969): 530–35.

44. Wells, *The Time Machine,* ed. Moorcock, 19.

45. Wells, *The Time Machine,* 85.

46. Adams, *Letter to American Teachers,* 145.

47. For the earlier history of comets as omens of reformations and catastrophes, see Sara Schechner Genuth, *Comets, Popular Culture, and the Birth of Modern Cosmology* (Princeton: Princeton University Press, 1997).

48. H. G. Wells, "The Star," *The Complete Short Stories of H. G. Wells* (New York: St. Martin's, 1970), 646.

49. Wells, "The Star," 654–55.

50. Zamyatin noted the Wells/Flammarion connection in "H. G. Wells," in *A Soviet Heretic,* 288. Flammarion's novelized work of scientific popularization was translated into a dozen languages, including the hasty 1894 English version. W. H. G. Armytage has observed that *La Fin du Monde* "must have influenced Wells though Wells never once mentioned Flammarion in his writings." W. H. G. Armytage, *Yesterday's Tomorrows: A Historical Survey of Future Societies* (Toronto: University of Toronto Press, 1968), 34. Perhaps Wells wished to keep under wraps the existence of this veritable Vulcan's workshop of science-fiction machinery.

51. Flammarion, *Omega,* 11, 12.

52. Flammarion, *Omega,* 57–58.

53. Later in the debate, another version of this thermodynamic scenario is aired: "If some star, active or extinct, should emerge from space, so as to form with the sun a sort of electro-dynamic couple with our planet on its axis, acting upon it like a brake—if, in a word, for any reason, the earth should be suddenly arrested in its orbit, its mechanical energy would be changed into molecular motion, and its temperature would be suddenly raised to such a degree as to reduce it entirely to a gaseous state" (Flammarion, *Omega* 113).

54. Wells, "The Star," 647.

55. See A. S. Eve and C. H. Creasey, *Life and Work of John Tyndall* (London: Macmillan, 1945); Charles S. Blinderman, "John Tyndall and the Victorian New Philosophy," *Bucknell Review* 9 (March 1961): 281–90; Paul L. Sawyer, "Ruskin and Tyndall: The Poetry of Matter and the Poetry of Spirit," in Paradis and Postlewait, *Victorian Science,* 217–46; Gillian Beer, "Helmholtz, Tyndall, Gerard Manley Hopkins: Leaps of the Prepared Imagination," *Comparative Criticism* 13 (1991): 117–45, and "Wave Theory" 206–9; and Thomas F. Gieryn, "John Tyndall's Double Boundary-Work: Science, Religion and Mechanics in Victorian England," in *Cultural Boundaries of Science: Credibility on the Line* (Chicago: University of Chicago Press, 1999), 37–64.

56. John Tyndall, "The Constitution of Nature" (1865), in *Fragments of Science: A Series of Detached Essays, Addresses, and Reviews,* 2 vols. (New York: D. Appleton, 1915), 1:6–7.

57. Flammarion, *Omega,* 107–10.

58. Flammarion, *Omega,* 217.

59. Flammarion, *Omega,* 197.

60. Flammarion, *Omega,* 202. Flammarion's notorious spiritualism breaks loose at this point and is largely suppressed by the English-language translator of *Omega.* In fact, Flammarion was transcribing a dialect of "aethereal vibrations" very closely tied to the late-classical description of light, heat, and electromagnetism as radiant energies transmitted by a luminiferous ether. His vision of intellectual evolution displaces into the marvelous future the physical scientisms at play in the rhetoric of fin de siècle spiritualism and theosophy. With the evolution of the psychic sense, "the ethereal vibrations that result from cerebral movements are transmitted by a transcendental magnetism that children themselves know how to use. Every thought excites in the brain a vibratory movement; such movement gives birth to ethereal waves; and if such waves encounter another mind in harmony with the first mind, they can convey to it the initial thought that created them, just as a vibrating wire receives the undulation from a distance sound, and as the speaker of a telephone reconstitutes the voice silently transported by an electrical movement" (*La Fin du Monde,* 306–8; my translation).

61. Flammarion, *Omega,* 272.

62. Flammarion, *Omega,* 275.

63. John Tyndall, *Heat: A Mode of Motion,* 6th ed. (1880; New York: D. Appleton, 1915), 523.

64. Tyndall, *Heat,* 519–20.

65. Tait and Stewart, *Unseen Universe,* 92.

66. Tyndall, "The Constitution of Nature," 7.

67. Flammarion, *Omega,* 285–87.

68. James Joyce, *Ulysses* (New York: Vintage, 1990), 700.

69. Zamyatin, *A Soviet Heretic,* 107.

70. Buck-Morss, *The Dialectics of Seeing,* 95.

Chapter 6

1. Zamyatin, *We,* 3.

2. Cf. Leighton Brett Cooke, "The Manuscript in Zamjatin's *We,*" *Russian Literature* 17 (1985): 367–88.

3. Zamyatin, *We,* 156.

4. Zamyatin, *We,* 9.

5. "Like Henry Adams, Zamjatin was sufficiently impressed by the forecast of universal heat death, popularized in H. G. Wells' *The Time Machine,* to apply the notion to his views on society and history." Efraim Sicher, "The Last Utopia: Entropy and Revolution in the Poetics of Evgeny Zamjatin," *History of European Ideas* 13, no. 3 (1991): 235.

6. See Booker, *Dystopian Literature,* 292–97; and Paul R. Josephson, *Totalitarian Science and Technology* (Atlantic Heights, N.J.: Humanities Press, 1996), 82.

7. Zamyatin, *We,* 22–23.

8. Hakfoort, "Science Deified," 525–44.

9. Hakfoort, "Science Deified," 536.

10. See chap. 4, n. 33 on Ostwald's copious output of *Sonntagspredigten* for the German Association of Monists.

11. Richard Stites, *Revolutionary Dreams: Utopian Vision and Experimental Life in the Russian Revolution* (New York: Oxford University Press, 1989), esp. "Man the Machine" (145–64).

12. "Taylor's goal was to use the worker's energy and time with minimal waste—no wasted seconds, no wasted motions. Two concerns were dominant: speed and regulation. The fastest and smoothest pattern of activity became the only pattern; individual variations were not tolerated." Carolyn H. Rhodes, "Frederick Winslow Taylor's System of Scientific Management in Zamyatin's *We*," *Journal of General Education* 28, no. 1 (1976): 32. Cf. Hubert Zapf, "Taylorism in D. H. Lawrence's *Women in Love*," *D. H. Lawrence Review* 15, nos. 1–2 (1982): 129–39. See also James F. Knapp, "The Discourse of Scientific Management," *Literary Modernism and the Transformation of Work* (Evanston, Ill.: Northwestern University Press, 1988), 1–18; Rabinbach, *The Human Motor,* 238–58 and passim; and Josephson, *Totalitarian Science and Technology,* 75–83.

13. Zamyatin, *We,* 5–6.

14. Stites, *Revolutionary Dreams,* 188.

15. Zamyatin, *We,* 172.

16. "In the author's projection of a world where the concept of private parenthood has been rejected, he refers to the impersonal process that has replaced parenthood as 'child production,' a phrase that sounds like a reference to an industrial process. A similar emphasis on commodity is evident in the *Lex Sexualis* which regulates the assignment of partners: 'A Number may obtain a license to use any other Number as a sexual product'" (Rhodes, "Taylor's System," 37).

17. Zamyatin, *We,* 7.

18. Zamyatin, *We,* 176.

19. Zamyatin, *We,* 80.

20. Zamyatin, *We,* 177.

21. Efraim Sicher characterizes the Biblical archetype of utopia as a refuge cloistered from care: "Utopias have characteristically been represented as closed spaces in which happiness is sought behind secure and forbidding walls" ("The Last Utopia," 226). See also Richard A. Gregg, "Two Adams and Eve in the Crystal Palace: Dostoevsky, the Bible and *We,*" in *Zamyatin's "We": A Collection of Critical Essays,* ed. Gary Kern (Ann Arbor, Mich.: Ardis, 1988), 61–69.

22. See Ian Hacking, "Probability and Determinism, 1650–1900," in *Companion to the History of Modern Science,* ed. R. C. Olby et al. (New York: Routledge, 1990), 690–701.

23. Prigogine and Stengers, *Order Out of Chaos,* 123.

24. Prigogine and Stengers, *Order Out of Chaos,* 123.

25. Prigogine and Stengers, *Order Out of Chaos,* 128.

26. Cf. S. A. Cowan, "The Crystalline Center of Zamyatin's *We,*" *Extrapolation* 29, no. 2 (1988): 160–78.

27. Prigogine and Stengers, *Order Out of Chaos,* 126.

28. Prigogine and Stengers, *Order Out of Chaos,* 127.

29. Coveney and Highfield, *Frontiers of Complexity,* 152.

30. Cf. Darko Suvin's verdict that "Zamyatin's novel is not consistent. Of a bold general concept, it hesitates midway between Chernyshevsky and Dostoevsky—undecided as to what it thinks of science and reason" (*Metamorphoses of Science Fiction,* 257).

31. Zamyatin, *We,* 159.

32. Zamyatin, *We,* 168.

33. Zamyatin, *A Soviet Heretic,* 107.

34. On dissipative structures, see Prigogine and Stengers, *Order Out of Chaos,* esp. chapter 5, "The Three Stages of Thermodynamics," 131–76. On the sociological extension of the concept of autopoiesis, see Niklas Luhmann, "The Autopoiesis of Social Systems," in *Essays on Self-Reference* (New York: Columbia University Press, 1990), 1–20.

35. For more on scientism in Lawrence and modernism generally, see my *Dora Marsden,* esp. 154–71.

36. D. H. Lawrence, *The Crown,* in *Reflections on the Death of a Porcupine and Other Essays,* ed. Michael Herbert (New York: Cambridge University Press, 1988), 255. Curiously, there is no mention of the parable of the cave in Barry J. Scherr, *D. H. Lawrence's Response to Plato: A Bloomian Interpretation* (New York: Peter Lang, 1996).

37. Lawrence, *The Crown,* 287.

38. See David Ellis, "Lawrence and the Biological Psyche," in *D. H. Lawrence: Centenary Essays,* ed. Mara Kalnins (Bristol: Bristol Classical Press, 1986), 89–109. Other important recent works on Lawrence's doctrinal writings include David Ellis and Howard Mills, *D. H. Lawrence's Non-Fiction: Art, Thought, and Genre* (New York: Cambridge University Press, 1988); Michael Black, *D. H. Lawrence: The Early Philosophical Works, a Commentary* (New York: Cambridge University Press, 1992); Anne Fernihough, *D. H. Lawrence: Aesthetics and Ideology* (New York: Oxford University Press, 1993); and Fiona Becket, *D. H. Lawrence: The Thinker as Poet* (New York: St. Martin's, 1997).

39. For illuminating summaries of the rise of modern physics and treatments of Lawrence's relation to those developments, see N. Katherine Hayles, "The Ambivalent Approach: D. H. Lawrence and the New Physics," *Mosaic* 15, no. 3 (1982): 89–108, and "Evasion: The Field of the Unconscious in D. H. Lawrence," in *The Cosmic Web: Scientific Field Models and Literary Strategies in the Twentieth Century* (Ithaca, N.Y.: Cornell University Press, 1984), 85–110.

40. Peter Harman points out that "appeal to theological arguments, not uncommon among British physicists, was a manifestation of the continued influence of the 'natural theology' tradition" (*Energy, Force, and Matter,* 68–69). In particular, William Thomson's universal moralization of mechanical dissipation was motivated by theological considerations. See Crosbie Smith, "Natural Philosophy and Ther-

modynamics: William Thomson and 'The Dynamical Theory of Heat,'" *British Journal for the History of Science* 9 (1976): 293–319.

41. For a provocative critique of Lawrence's complex attitudes toward visuality, see Linda Ruth Williams, *Sex in the Head: Visions of Femininity and Film in D. H. Lawrence* (Detroit: Wayne State University Press, 1993).

42. Lawrence, *The Crown,* 287.

43. Cited in Mabel Dodge Luhan, *Lorenzo in Taos* (London: Martin Secker, 1933), 21–23: "Mathematics is founded on the void, and therefore escapes from the void; the optic nerve is an organized light beam, the brain an organized sun, the eye an organized spectrum or rainbow."

44. D. H. Lawrence, *Psychoanalysis and the Unconscious* and *Fantasia of the Unconscious* (New York: Viking, 1976), 55.

45. Lawrence, *Fantasia of the Unconscious,* 63–64.

46. For Edward Carpenter on the life force, see "Civilisation: Its Cause and Cure," in *Civilisation,* 12th ed., 1–50. I discuss this text as a precursor of modernist vitalism in *Dora Marsden,* 32–38. On Lawrence and Carpenter, see Emile Delavenay, *D. H. Lawrence and Edward Carpenter: A Study in Edwardian Transition* (New York: Taplinger, 1971).

47. Tait and Stewart, *Unseen Universe,* 176–77. The *Unseen Universe* held the concepts of energy and life apart long enough to reinscribe them both in a personification allegory, as different Persons of the Christian Trinity: "The Third Person of the Trinity is regarded in this system as working in the universe, and therefore in some sense as conditioned, and as distributing and developing this principle of life, which we are forced to regard as one of the things of the universe, in the same manner as the second person of the Trinity is regarded as developing that other phenomenon, the energy of the universe" (179).

48. Lawrence, *Fantasia of the Unconscious,* 57.

49. Lawrence, *Fantasia of the Unconscious,* 56–57.

50. For a quick overview of German Romantic science, see David Knight, *The Age of Science: The Scientific World-View in the Nineteenth Century* (Oxford: Basil Blackwell, 1986), 56–57. See also Guenter B. Risse, "Kant, Schelling, and the Early Search for Philosophical 'Science' of Medicine in Germany," *Journal of the History of Medicine and Allied Sciences* 27 (1972): 145–58; and Karl E. Rothschuh, "The So-Called 'Romantic' Interlude in Physiology," in *History of Physiology,* trans. Guenter B. Risse (Huntington, N.Y.: R. E. Krieger, 1973), 155–66. On electrobiology, see Alfred Smee, *Elements of Electro-Biology, or, The Voltaic Mechanism of Man; of Electro-Pathology, especially of the Nervous system; and of Electro-Therapeutics* (London: Longman, Brown, 1849).

51. D. H. Lawrence, *Aaron's Rod* (New York: Penguin, 1976), 346.

52. Lawrence, *Fantasia of the Unconscious,* 215.

53. D. H. Lawrence, *Lady Chatterley's Lover* (New York: Modern Library, 1959).

54. Lawrence, *Lady Chatterley's Lover,* 264.

55. Lawrence, *Lady Chatterley's Lover,* 266.
56. Lawrence, *Lady Chatterley's Lover,* 343.

Chapter 7

1. D. W. Theobald has noted "the vigorous positivism of many philosophers of science towards the end of the nineteenth century. . . . Names like Mach and Pearson immediately spring to mind. Many commentators at that time inclined to the view that scientists were conceiving too much and experimenting too little, and that theory and observation were in danger of becoming, if they had not already become, too tenuously connected. Nineteenth-century science developed a body of theory which many found difficult to believe was a description of physical reality at all, and by way of reaction it became customary to regard science as basically concerned with ordering observations" (*The Concept of Energy,* 21). In the next chapter we will examine similar opinions in J. B. Stallo's *The Concepts and Theories of Modern Physics* (1882).

2. "Ether provided a means for resolving apparent discontinuities in the spatial-material order, and even for resolving the fundamental discontinuity between material and non-material orders. This latter possibility led some of the scientists into extra-scientific speculation, thus implying the character of ether as a general cultural fiction" (Benson, "Facts and Fictions," 830).

3. See Frank Wilczek, "The Persistence of Ether," *Physics Today,* January 1999, 11–13: "Einstein first purified, then enthroned, the ether concept. As the 20th century has progressed, its role in fundamental physics has only expanded. At present, renamed and thinly disguised, it dominates the accepted laws of physics. . . . [In] Einstein's original paper on special relativity . . . the Faraday-Maxwell concept of electric and magnetic fields, as media or ethers filling all space, was retained. What had to be sacrificed was only the false intuition that motion at a constant velocity would necessarily modify the equations of an ether" (11). Thanks to Luis Arata for this reference.

4. See Cantor and Hodge, *Conceptions of Ether,* 1–2.

5. During the eighteenth century, "light was frequently regarded as analogous to 'fire' and represented as an imponderable fluid 'ether.' These 'fluids' were envisaged as composed of mutually repelling particles (this property being denoted as 'elasticity'), as being 'subtle' (being able to penetrate the empty spaces between the particles of ordinary matter), and as being attracted by ordinary matter" (Harman, *Energy, Force, and Matter,* 13–14).

6. James Clerk Maxwell, "Ether," *Scientific Papers,* ed. Niven, 2:763.

7. Gillispie states that "the two great aspects of nineteenth-century physics—thermodynamics and mechanics . . . both evolved in similar patterns: exorcising by a sophisticated statement a hypostatic fluid constructed for spatial transmission." The two hypothetical substances in question here are caloric and ether. "Caloric vanished into a more elaborate energetics, of course. Nevertheless, as a construct it was the starting point of thermodynamics. . . . Aether accommodated more funda-

mental relations than caloric, and was correspondingly more difficult to exorcise; but as caloric disappeared into thermodynamics, so aether did into relativity" (*The Edge of Objectivity,* 356–57).

8. Martin Goldman drives this point home: "In thermodynamics, the 2nd law (or its equivalent), which was undoubtedly correct, had been derived at least twice on the basis of models that were undoubtedly incorrect: Rankine had done it for molecular vortices, and earlier the Frenchman Sadi Carnot had discovered the law on the basis of the Caloric fluid theory of heat" (*Demon in the Aether,* 154).

9. Benson, "Facts and Fictions," 828.

10. Cited in Robert S. Elliott, *Electromagnetics: History, Theory, and Applications* (Piscataway, N.J.: IEEE Press, 1993), 38.

11. Maxwell, "Ether," 2:768.

12. Maxwell, "Ether," 2:775.

13. See Elliott, *Electromagnetics,* 39–40; and Berkson, *Fields of Force,* 265–75.

14. In Benson's succinct narration of these developments, by the turn of the twentieth century, "the ether fiction had very nearly exhausted its usefulness and hence its credibility. Field theory had filled space with fields of force (and, in a sense, matter) and was redirecting physics toward purely abstract mathematical analysis of energy distributions without regard to a medium. Planck's reduction of radiant energy to quanta and Einstein's reduction of light to photons would shortly mark the formal acceptance of fundamental discontinuities into physical description. And relativity theory, by denying absolute rest and thus absolute measurement of motion, would eliminate the theoretical necessity for any absolute spatial reference" ("Facts and Fictions," 836).

15. Beer, "Helmholtz, Tyndall, Hopkins," 121.

16. Beer, "Helmholtz, Tyndall, Hopkins," 122.

17. Tyndall, "The Constitution of Nature," 1:4.

18. John Tyndall, "Lecture X," in *Heat,* 276. See also Stephen G. Brush, "The Wave Theory of Heat: A Forgotten Stage in the Transition from the Caloric Theory to Thermodynamics," *British Journal for the History of Science* 5, no. 18 (1970): 145–67.

19. John Tyndall, "Radiation" (1865), in *Fragments of Science,* 1:34–35.

20. Tyndall, "Radiation," 1:72–73.

21. John Tyndall, "On Radiant Heat in Relation to the Colour and Chemical Constitution of Bodies" (1866), in *Fragments of Science,* 1:76.

22. Tyndall, "On Radiant Heat," 1:76–77.

23. John Tyndall, "The Scientific Use of the Imagination," in *Fragments of Science,* 2:103.

24. Tait and Stewart, *Unseen Universe,* 116.

25. Maxwell, *Scientific Papers,* ed. Niven, 2:758.

26. Tait and Stewart, *Unseen Universe,* 156.

27. Tait and Stewart, *Unseen Universe,* 158.

28. On allegory as a rhetorical mode of doctrinal defense, see Clarke, *Allegories of Writing,* 11, 42, and passim.

29. Beer, "Helmholtz, Tyndall, Hopkins," 121.

30. H. P. Blavatsky, *Isis Unveiled: A Master-Key to the Mysteries of Ancient and Modern Science and Theology,* vol. 1, *Science* (1877; rpt. Covina, Calif.: Theosophical University Press, n.d.).

31. See Linda D. Henderson, "Mysticism, Romanticism, and the Fourth Dimension," in *The Spiritual in Art: Abstract Painting, 1890–1985* (New York: Abbeville, 1986), 222.

32. Blavatsky, *Isis Unveiled,* 129.

33. Blavatsky, *Isis Unveiled,* 129.

34. Blavatsky, *Isis Unveiled,* 134.

35. See Peter Washington, *Madame Blavatsky's Baboon: A History of the Mystics, Mediums, and Misfits Who Brought Spiritualism to America* (New York: Schocken, 1995).

36. See Stephen W. Hawking, *A Brief History of Time* (New York: Bantam, 1988), 25; and Roger Penrose, *The Emperor's New Mind: Concerning Computers, Minds, and the Laws of Physics* (New York: Oxford University Press, 1990), 193–221.

37. Tait and Stewart, *Unseen Universe,* 156.

38. Blavatsky, *Isis Unveiled,* 180.

39. Robbin, *Fourfield,* 24–38.

40. Tait and Stewart, *Unseen Universe,* 221.

41. Hinton, "What is the Fourth Dimension?" 31.

42. Tyndall, "The Constitution of Nature," 7.

43. Hinton, *New Era of Thought,* 52.

44. See Tyndall on the law of inverse squares in *Heat,* 293–96; cf. Hinton, appendix to "A Plane World," 1:157–59.

45. Hinton, "A Picture of Our Universe," in *Scientific Romances,* 1:192.

46. Hinton, "Picture of Our Universe," 1:194.

47. *Timaeus,* 28b–c. See Francis McDonald Cornford, *Plato's Cosmology: The "Timaeus" of Plato Translated with a Running Commentary* (London: Routledge and Kegan Paul, 1956), 23.

48. Hinton, "What is the Fourth Dimension?" 4.

49. Hinton, "What is the Fourth Dimension?" 13.

50. Hinton, "Many Dimensions," in *Speculations on Fourth Dimension,* ed. Rucker, 78.

51. Hinton, "Many Dimensions," 18.

52. Hinton, "Many Dimensions," 25.

53. Hinton, "Many Dimensions," 18.

54. Cited in Kittler, *Gramophone, Film, Typewriter,* 30–31.

55. Hinton, "Picture of Our Universe," 1:196–97.

56. Hinton, "Picture of Our Universe," 1:197.

57. Kittler, *Gramophone, Film, Typewriter,* 33.

58. Kittler, *Gramophone, Film, Typewriter,* 12.

59. Cited in Kittler, *Gramophone, Film, Typewriter,* 40.

60. Kittler, *Gramophone, Film, Typewriter,* 85.

61. Cf. Myers, "Nineteenth-Century Popularizations": "For Stewart and Tait 'personality' is at the center of physics and motivates all important research" (331).

62. Hinton, *New Era of Thought,* 65.

63. Hinton, *New Era of Thought,* 66.

64. "Appended to the essay 'Hypercubes' in Martin Gardner's book *Mathematical Carnival,* there is a curious letter from a former user of Hinton's cubes who calls them 'completely mind-destroying'" (Rucker, in Hinton, *Speculations on Fourth Dimension,* xv).

65. The philosophical discourse of the fourth dimension in the later modernist period is more often associated with P. D. Ouspensky's *Tertium Organum.* However, concepts of a spatial fourth dimension had already made a considerable impact on modernist doctrines well before translations of the Russian original became generally available in the 1920s. Moreover, Ouspensky gave ample credit to Hinton as a pioneer on the topic. The first English edition is P. D. Ouspensky, *Tertium Organum: A Key to the Enigmas of the World,* trans. Nicholas Bessaraboff and Claude Bragdon (Rochester, N.Y.: Manas Press, 1920). Bragdon's *A Primer of Higher Space (The Fourth Dimension)* (Rochester, N.Y.: Manas Press, 1913) anticipates Ouspensky by synthesizing Hinton's discourses with a theosophical approach to human evolution.

66. Mabel Dodge Luhan, *Movers and Shakers* (New York: Harcourt, Brace, 1936), 39. See Lois Palken Rudnick, *Mabel Dodge Luhan: New Woman, New Worlds* (Albuquerque: University of New Mexico Press, 1984), chap. 3.

67. Luhan, *Movers and Shakers,* 39. See Linda Dalrymple Henderson, "Mabel Dodge, Gertrude Stein, and Max Weber: A Four-Dimensional Trio," *Arts,* September 1982, 106–11. For a wide view of scholarship on visionary modernism, see Anthony Trembley and Archie Henderson, "Modernism and the Occult: A Selected Bibliography," in *Literary Modernism and the Occult Tradition,* ed. Leon Surette and Demetres Tryphonopoulos (Orono, Maine: National Poetry Foundation, 1996), 1–16.

68. "The idea of the universe as waves, of the parallels between light, heat, and sound, and the single process expressed through them, enters late-nineteenth-century writing with a fresh urgency. Flux, the vortex, the ocean, the aura, the 'sea of forces flowing and rushing together,' as Nietzsche called it, so important in modernism, are all elements of a repertoire shifting across fields" (Beer, "Wave Theory," 209); see also Linda Dalrymple Henderson, "Modern Art and the Invisible: The Unseen Waves and Dimensions of Occultism and Science," in *Okkultismus und Avantgarde 1900–1915* (Frankfurt am Main: Schirn Kunsthalle Frankfurt, 1995), and "Duchamp and Invisible Reality," *Duchamp in Context,* 3–28.

69. Blavatsky, *Isis Unveiled,* 126.

70. Edward Carpenter, *The Art of Creation: Essays on the Self and Its Powers* (London: George Allen, 1907), 33–34. See my *Dora Marsden,* 32–38 and passim; and Linda Dalrymple Henderson, "Mysticism as the 'Tie That Binds': The Case of Edward Carpenter and Modernism," *Art Journal* 46 (spring 1987): 29–37.

71. See Erik Davis, "The Alchemical Fire," *Techgnosis,* 39–75; and Gillian Beer, " 'Wireless': Popular Physics, Radio, and Modernism," in *Cultural Babbage: Technology, Time, and Invention,* ed. Francis Spufford and Jenny Uglow (Boston: Faber and Faber, 1996), 149–66.

72. For the spiritualist slant on the ether, see Janet Oppenheim, *The Other World: Spiritualism and Psychical Research in England, 1850–1914* (New York: Cambridge University Press, 1985), 370–87. It is curious that in Alison Winter's *Mesmerized: Powers of Mind in Victorian Britain* (Chicago: University of Chicago Press, 1998), the discussion of "ether" revolves around the medicinal distillate of that name—hypnosis and ether being rival modes of surgical anaesthesia—while the ether of space is nowhere mentioned.

73. Mesmer wrote: "For a long time I have supposed that a universal fluid exists in nature, a fluid which penetrates all animate or inanimate bodies. The phenomena of electricity, as well as those of magnetism, affected this opinion of mine. Thus, I adopted the system of the noble Newton regarding the motion of celestial bodies." Franz Anton Mesmer, *Mesmerism: A Translation of the Original Scientific and Medical Writings of F. A. Mesmer,* trans. George Bloch (Los Altos, Calif.: William Kaufmann, 1980), 33. See also Alan Gauld, *A History of Hypnotism* (New York: Cambridge University Press, 1992); and Adam Crabtree, *From Mesmer to Freud: Magnetic Sleep and the Roots of Psychological Healing* (New Haven: Yale University Press, 1993).

74. Mesmer, *Mesmerism,* 43.

75. Mesmer, *Mesmerism,* 121.

76. For a cranky treatment of Lawrence's debt to the Mesmeric tradition, see Maria M. Tatar, *Spellbound: Studies on Mesmerism and Literature* (Princeton: Princeton University Press, 1978), 243–55.

77. Luhan, *Lorenzo in Taos,* 43.

78. His brief sketch beginning a projected novelization of Mabel's life, quickly aborted by Frieda's disapprobation, was entitled "The Wilful Woman." The spiritual condition of Mabel's fictive persona, Sybil Mond, as she languished on the slow train through the Sangre de Cristo Mountains is brought forward with a nice electromagnetic flourish: "Volts of distracted impatience and heart-brokenness surged out of her" (*Lawrence and New Mexico,* 7).

79. Luhan, *Lorenzo in Taos,* 118; rpt. in *The Letters of D. H. Lawrence,* 7 vols. (New York: Cambridge University Press, 1979–93), 4:514. I cite the letters from *Lorenzo in Taos.*

80. D. H. Lawrence, *Women in Love* (New York: Viking, 1975), 235.

81. D. H. Lawrence, *The Symbolic Meaning: The Uncollected Versions of "Studies in Classic American Literature"* (New York: Viking, 1964), 24–25.

82. Lawrence, *Symbolic Meaning,* 162.

83. See Prigogine and Stengers, *Order Out of Chaos,* 79–85.

84. Lawrence, *Symbolic Meaning,* 162–63.

85. D. H. Lawrence, *Birds, Beasts, and Flowers!* (Santa Rosa, Calif.: Black Sparrow Press, 1992), 50–51.

86. D. H. Lawrence, "Indians and Entertainment," in *Morning in Mexico,* ed. Ross Parmenter (Salt Lake City: Gibbs M. Smith, 1982), 109.

Chapter 8

1. Albert Einstein, *Relativity: The Special and General Theory,* trans. Robert W. Lawson (1920; New York: Crown, 1961), 16.
2. Einstein, *Relativity,* 10.
3. Elliott, *Electromagnetics,* 38.
4. Einstein, *Relativity,* 59.
5. Einstein, *Relativity,* 21.
6. Einstein, *Relativity,* 41.
7. Sir Oliver Lodge, *Ether and Reality: A Series of Discourses on the Many Functions of the Ether of Space* (London: Hodder and Stoughton, 1925), and *My Philosophy, Representing My Views on the Many Functions of the Ether of Space* (London: E. Benn, 1933). See Oppenheim, *The Other World,* 371–97.
8. Unless otherwise indicated, quotations from Carpenter's "Modern Science: A Criticism" in this chapter are from *Civilisation: Its Cause and Cure and Other Essays,* 2d ed. (London: Swan Sonnenschein, 1891), 51–81.
9. Letter from Edward Carpenter to Havelock Ellis, May 22, 1920, Harry Ransom Humanities Research Center, University of Texas.
10. Carpenter, "Modern Science: A Criticism," 51.
11. Sir James Jeans, *The Mysterious Universe* (Cambridge: Cambridge University Press, 1930), 108–9.
12. Carpenter, "Modern Science: A Criticism," 70.
13. Edward Carpenter, *Civilization: Its Cause and Cure and Other Essays,* 14th ed. (New York: Charles Scribner's Sons, 1921), 88. Cf. Edward Carpenter, "Modern Science: a criticism," page proofs with A revisions and 1 p. Ams, December 17, 1920, Harry Ransom Humanities Research Center, University of Texas.
14. "There is yet another idea of science that we feel is potentially just as detrimental, namely, the fascination with a mysterious science that, by paths of reasoning inaccessible to common mortals, will lead to results that can, in one fell swoop, challenge the meaning of basic concepts such as time, space, causality, mind, or matter. This kind of 'mystery science,' the results of which are imagined to be capable of shattering the framework of any traditional conception, has actually been encouraged by the successive 'revelations' of relativity and quantum mechanics" (Prigogine and Stengers, *Order Out of Chaos,* 34).
15. Carpenter, "Modern Science: A Criticism," *Civilisation,* 14th ed., 71–72.
16. Stallo, *Concepts and Theories,* 128. See Albert Moyer, *American Physics in Transition: A History of Conceptual Change in the Late Nineteenth Century* (Los Angeles: Tomash, 1986), 3–32; thanks to Bruce Hunt for this latter reference.
17. Stallo, *Concepts and Theories,* 247; cf. Carpenter, *Civilisation,* 2d ed., 72.
18. Stallo, *Concepts and Theories,* 209–10.
19. "In the early years of Bolshevik rule the theory of relativity provided a bat-

tlefield for a fierce struggle between those members of the intelligentsia who espoused an orthodox materialist outlook and those who argued from an idealist position": W. J. Leatherbarrow, "Einstein and the Art of Yevgeny Zamyatin," *MLR* 82, no. 1 (1987): 144. See also David Joravsky, *Soviet Marxism and Natural Science, 1917–1932* (New York: Columbia University Press, 1961).

20. Zamyatin, "The New Russian Prose," in *A Soviet Heretic,* 105.

21. "Various statements made by I-330, one of the leaders of the rebellious MEFIS, clearly demonstrate that those who oppose the state have a much deeper understanding of mathematics and its empirical implications than those who support the state. . . . It is as if science and mathematical thought have been largely defunct for the past thousand years." Leighton Brett Cooke, "Ancient and Modern Mathematics in Zamyatin's *We,*" in Kern, *Zamyatin's "We,"* 151.

22. Zamyatin, *A Soviet Heretic,* 87.

23. See Linda Dalrymple Henderson, "Transcending the Present: The Fourth Dimension in the Philosophy of Ouspensky and in Russian Futurism and Suprematism," in *The Fourth Dimension,* 238–99.

24. Zamyatin, "H. G. Wells," 285.

25. Zamyatin, "On Synthetism," 86.

26. Zamyatin, "Literature, Revolution, Entropy," 107.

27. Zamyatin, *We,* 21.

28. Zamyatin, *We,* 3, 4. Integration in mathematics is defined as "the process of finding the equation or function of which a given quantity or function is the derivative." In *Order Out of Chaos,* Prigogine and Stengers remark: "Any integrable system may . . . be represented as a set of units, each changing in isolation, quite independently of all the others, in that eternal and immutable motion Aristotle attributed to the heavenly bodies"; Poincaré's proof of the insolubility of the three-body problem "announced the end of all simple extrapolations of dynamics based on integrable systems" (71–72).

29. Zamyatin, *We,* 7.

30. Zamyatin, *We,* 10.

31. Zamyatin, *We,* 27, 41.

32. Zamyatin, *We,* 91.

33. Zamyatin, *We,* 84. Brett Cooke makes the fourth-dimensional context of this detail explicit in his commentary on the mathematics of the novel: "D-503 continually gets only fleeting glimpses of S-4711, who pops in and out of sight as if this Guardian and covert member of the MEFIS were a four-dimensional being intruding on D-503's three-dimensional realm" ("Mathematics in *We,*" 158).

34. Zamyatin, *We,* 86.

35. Zamyatin, *We,* 87.

36. Zamyatin, *We,* 90.

37. Zamyatin, *We,* 98.

38. Cooke, "Mathematics in *We,*" 159.

39. Zamyatin, "On Synthetism," 81.

40. "Spirals figure prominently in Zamyatin's essays where he uses them to rep-

resent paths of escape from the 'dead circles' of dogma and artistic convention, such as are embodied in the Single State. Zamyatin infers that it is a simple matter to escape from a circle: all one has to do is to enter an extra dimension" (Cooke, "Mathematics in *We,*" 159).

41. Bragdon, *Primer of Higher Space,* 40, plate 16. See also Hinton, *The Fourth Dimension* (1904), 25; and Henderson, *Fourth Dimension* 158–59, and plate 32. Thanks for other assistance to Linda D. Henderson and Ian F. A. Bell.

42. Lawrence, *Fantasia of the Unconscious,* 208.

43. Lawrence's uncharacteristically receptive attitude to Einstein is noted in William York Tindall, *D. H. Lawrence and Susan His Cow* (New York: Columbia University Press, 1939), 31; David Ellis, "Poetry and Science in the Psychology Books," in Ellis and Mills, *D. H. Lawrence's Non-Fiction,* 77; and Mark Kinkead-Weekes, *D. H. Lawrence: Triumph to Exile, 1912–1922* (New York: Cambridge University Press, 1996), 659.

44. Lawrence, *Fantasia of the Unconscious,* 66.

45. Lawrence, *Fantasia of the Unconscious,* 208.

46. Lawrence, *Fantasia of the Unconscious,* 181.

47. Lawrence, *Fantasia of the Unconscious,* 66.

48. Einstein, *Relativity,* 84.

49. Lawrence, *Fantasia of the Unconscious,* 67.

50. Lawrence, *Fantasia of the Unconscious,* 180.

51. Lawrence, *Fantasia of the Unconscious,* 209.

52. Lawrence, *Fantasia of the Unconscious,* 66.

53. See E. W. Tedlock Jr., "D. H. Lawrence's Annotations of Ouspensky's *Tertium Organum,*" *Texas Studies in Literature and Language* 2, no. 2 (1960): 206–18; and Christopher Pollnitz, "D. H. Lawrence's Quarrel with P. D. Ouspensky: Which Door to the Fourth Dimension?" in *Modernism's Fourth Dimensions,* ed. Bruce Clarke and Linda Dalrymple Henderson, forthcoming from Pennsylvania State University Press.

54. Lawrence, *Fantasia of the Unconscious,* 180–81.

55. Lawrence, *The Crown,* 287.

56. Lawrence, "Why the Novel Matters," in *Study of Thomas Hardy and Other Essays,* ed. Bruce Steele (New York: Cambridge University Press, 1985), 194.

57. Lawrence, "Morality and the Novel," in *Study of Thomas Hardy,* 171.

58. Carpenter, *The Art of Creation,* 7.

59. Richard O. Young, " 'Where Even the Trees Come and Go': D. H. Lawrence and the Fourth Dimension," *D. H. Lawrence Review* 13, no. 1 (1980): 31. For an impressionistic adaptation of Young's scholarship to a novella written contemporaneously with the essays developing Lawrence's version of the fourth dimension, see Peter Balbert, "Scorched Ego, the Novel, and the Beast: Patterns of Fourth Dimensionality in *The Virgin and the Gypsy,*" *Papers on Literature and Language* 29, no. 4 (1993): 395–416.

60. Young cites a letter of 1908 to Blanche Jennings in which Lawrence had already coordinated the fourth dimension with Life by supplementing its physical

and geometrical qualities with an ethereal function as a vital medium of soul: "Life has four dimensions," Lawrence wrote Jennings: "people's souls come flowing round me and touching me, and I feel them" (cited in Young, "Even the Trees," 31).

61. See Henderson, *Fourth Dimension,* esp. "Cubism and the New Geometries," 44–116.

62. Lawrence, *The Crown,* 250, 249.

63. D. H. Lawrence, "Reflections on the Death of a Porcupine," in *Reflections,* 358.

64. Lawrence, "Morality and the Novel," 171.

65. Henderson, *Fourth Dimension,* 162; see also Henderson, *Duchamp in Context,* 207–14.

66. Marcel Duchamp, *Marcel Duchamp: Notes,* ed. and trans. Paul Matisse (Boston: G. K. Hall, 1983).

67. Cited in Henderson, *Fourth Dimension,* 75.

68. Lawrence, "Him with His Tail in His Mouth," in *Reflections,* 309.

69. Lawrence, "Introduction to These Paintings," in *Phoenix: The Posthumous Papers of D. H. Lawrence,* ed. Edward D. McDonald (1936; New York: Viking, 1968), 574.

70. Lawrence, "Him with His Tail," 310.

71. Lawrence, "Him with His Tail," 311.

72. Lawrence, "Him with His Tail," 313.

73. Lawrence, "Him with His Tail," 315–16.

Bibliography

Abbott, Edwin A. *Flatland: A Romance of Many Dimensions* (1886). New York: Dover, 1992.

Abernethy, Peter L. "Entropy in Pynchon's *The Crying of Lot 49*." *Critique* 14, no. 2 (1974): 18–32.

Adams, Henry. "A Letter to American Teachers of History" (1910). In *The Degradation of the Democratic Dogma*. New York: Macmillan, 1920.

Albright, Daniel. *Quantum Poetics: Yeats, Pound, Eliot, and the Science of Modernism*. Cambridge: Cambridge University Press, 1997.

Aldiss, Brian W. *The Billion Year Spree*. New York: Doubleday, 1973.

Annas, Julia. "Understanding and the Good: Sun, Line, and Cave." In *Plato's "Republic": Critical Essays*, ed. Richard Kraut, 143–68. New York: Rowman and Littlefield, 1997.

Aristotle, *Poetics*. In *The Critical Tradition: Classic Texts and Contemporary Trends*, ed. David H. Richter, 42–65. New York: St. Martin's, 1989.

Armytage, W. H. G. *Yesterday's Tomorrows: A Historical Survey of Future Societies*. Toronto: University of Toronto Press, 1968.

Asendorf, Christoph. *Batteries of Life: On the History of Things and Their Perception in Modernity*. Trans. Don Reneau. Berkeley and Los Angeles: University of California Press, 1993.

Assad, Maria. "Michel Serres: In Search of a Tropography." In *Chaos and Order: Complex Dynamics in Literature and Science*, ed. N. Katherine Hayles, 278–98. Chicago: University of Chicago Press, 1991.

Balbert, Peter. "Scorched Ego, the Novel, and the Beast: Patterns of Fourth Dimensionality in *The Virgin and the Gypsy*." *Papers on Literature and Language* 29, no. 4 (1993): 395–416.

Ballard, Marvin H. "The Life and Thought of Charles Howard Hinton." M.A. thesis, Virginia Polytechnic Institute, 1980.

Barker, Andrew, and Martin Warner, ed. *The Language of the Cave*. Edmonton: Academic, 1992.

Barney, Stephen A. *Allegories of History, Allegories of Love*. Hamden, Conn.: Archon Books, 1978.

Bateson, Gregory. *Steps to an Ecology of Mind*. New York: Ballantine, 1972.

Becket, Fiona. *D. H. Lawrence: The Thinker as Poet*. New York: St. Martin's, 1997.

Beehler, Michael. "Yevgeny Zamyatin: The Essential, the Superfluous, and Textual Noise." *SubStance* 15, no. 2 (1986): 48–60.

Beer, Gillian. "'The Death of the Sun': Victorian Solar Physics and Solar Myth." In *The Sun Is God: Painting, Literature, and Mythology in the Nineteenth Century*, ed. J. B. Bullen, 159–80. Oxford: Clarendon Press, 1989.

———. "Helmholtz, Tyndall, Gerard Manley Hopkins: Leaps of the Prepared Imagination." *Comparative Criticism* 13 (1991): 117–45.

———. "Wave Theory and the Rise of Literary Modernism." In *Realism and Representation: Essays on the Problem of Realism in Relation to Science, Literature, and Culture*, ed. George Levine, 193–213. Madison: University of Wisconsin Press, 1993.

———. "'Wireless': Popular Physics, Radio, and Modernism." In *Cultural Babbage: Technology, Time, and Invention*, ed. Francis Spufford and Jenny Uglow, 149–66. Boston: Faber and Faber, 1996.

Bell, Ian F. A. *Critic as Scientist: The Modernist Poetics of Ezra Pound*. New York: Methuen, 1981.

Bell, Michael. *Literature, Modernism, and Myth: Belief and Responsibility in the Twentieth Century*. New York: Cambridge University Press, 1997.

Benardete, Seth. *Socrates' Second Sailing: On Plato's "Republic."* Chicago: University of Chicago Press, 1989.

Benjamin, Walter. *The Arcades Project*. Trans. Howard Eiland and Kevin McLaughlin. Cambridge: Harvard University Press, 1999.

———. *The Origin of German Tragic Drama*. Trans. John Osborne. New York: Verso, 1985.

Bennett, Charles H. "Demons, Engines, and the Second Law." *Scientific American* 257 (1987): 108–16.

Benson, Donald R. "'Catching Light': Physics and Art in Walter Pater's Cultural Context." In *One Culture: Essays in Science and Literature*, ed. George Levine and Alan Rauch, 143–63. Madison: University of Wisconsin Press, 1987.

———. "Facts and Fictions in Scientific Discourse: The Case of Ether." *Georgia Review* 38 (1984): 825–37.

Berkson, William. *Fields of Force: The Development of a World View from Faraday to Einstein*. New York: John Wiley and Sons, 1974.

Birch, Dinah. "'The Sun Is God': Ruskin's Solar Mythology." In *The Sun is God: Painting, Literature, and Mythology in the Nineteenth Century*, ed. J. B. Bullen, 109–23. Oxford: Clarendon Press, 1989.

Black, Michael. *D. H. Lawrence: The Early Philosophical Works, a Commentary*. New York: Cambridge University Press, 1992.

Blavatsky, H. P. *Isis Unveiled: A Master-Key to the Mysteries of Ancient and Modern Science and Theology*. Vol. 1: *Science*. 1877; rpt. Covina, Calif.: Theosophical University Press, n.d.

Blinderman, Charles S. "John Tyndall and the Victorian New Philosophy." *Bucknell Review* 9 (March 1961): 281–90.

Bono, James J. "Science, Discourse, and Literature: The Role/Rule of Metaphor in Science." In *Literature and Science: Theory and Practice*, ed. Stuart Peterfreund, 59–89. Boston: Northeastern University Press, 1990.

Booker, M. Keith. *Dystopian Literature: A Theory and Research Guide.* Westport, Conn.: Greenwood Press, 1994.

Bork, Alfred M. "The Fourth Dimension in Nineteenth-Century Physics." *Isis* 55, no. 3 (1964): 328–30.

Born, Max. *Einstein's Theory of Relativity* (1924). New York: Dover, 1962.

Bourdieu, Pierre. *The Field of Cultural Production: Essays on Art and Literature,* ed. and intro. Randal Johnson. New York: Columbia University Press, 1993.

Braden, Charles S. *Spirits in Rebellion: The Rise and Development of New Thought.* Dallas: Southern Methodist University Press, 1963.

Bragdon, Charles. *A Primer of Higher Space (The Fourth Dimension).* Rochester, N.Y.: Manas Press, 1913.

Brantlinger, Patrick, ed. *Energy and Entropy: Science and Culture in Victorian Britain.* Bloomington: Indiana University Press, 1989.

Brown, Clarence. "Introduction: Zamyatin and the Persian Rooster." In *We,* by Yevgeny Zamyatin, trans. Clarence Brown, xiii–xxx. New York: Penguin, 1993.

Brush, Stephen G. *The Temperature of History: Phases of Science and Culture in the Nineteenth Century.* New York: Burt Franklin, 1978.

———. "The Wave Theory of Heat: A Forgotten Stage in the Transition from the Caloric Theory to Thermodynamics." *British Journal for the History of Science* 5, no. 18 (1970): 145–67.

Buck-Morss, Susan. *The Dialectics of Seeing: Walter Benjamin and the Arcades Project.* Cambridge: MIT Press, 1988.

Bukatman, Scott. "Amidst These Fields of Data: Allegory, Rhetoric, and the Paraspace." *Critique: Studies in Contemporary Fiction* 33, no. 3 (1992): 199–219.

Bullen, J. B., ed. *The Sun Is God: Painting, Literature, and Mythology in the Nineteenth Century.* Oxford: Clarendon Press, 1989.

Bulwer-Lytton, Edward. *The Coming Race, Falkland, Zicci,* and *Pausanias the Spartan.* London: Routledge, n.d.

Cameron, Iain, and David O. Edge. *Scientific Images and Their Social Uses: An Introduction to the Concept of Scientism.* Boston: Butterworth, 1979.

Campbell, James L. "Edward Bulwer-Lytton's *The Coming Race* as a Condemnation of Advanced Ideas." *Essays in Arts and Sciences* 16 (May 1987): 55–63.

Campbell, Lewis, and William Garnett. *The Life of James Clerk Maxwell.* London: Macmillan, 1882.

Cantor, G. N., and M. J. S. Hodge, eds. *Conceptions of Ether: Studies in the History of Ether Theories, 1740–1900.* New York: Cambridge University Press, 1981.

Carpenter, Edward. *The Art of Creation: Essays on the Self and Its Powers.* London: George Allen, 1907.

———. *Civilisation: Its Cause and Cure and Other Essays.* 2d ed. London: Swan Sonnenschein, 1891.

———. *Civilisation: Its Cause and Cure and Other Essays.* 12th ed. London: George Allen, 1912.

———. *Civilization: Its Cause and Cure and Other Essays.* 14th ed. New York: Charles Scribner's Sons, 1921.

———. Letter to Havelock Ellis, May 22, 1920. Harry Ransom Humanities Research Center, University of Texas.

Cassirer, Ernst. *Determinism and Indeterminism in Modern Physics.* New Haven: Yale University Press, 1956.

Chamberlain, J. Edward. "Images of Degeneration: Turnings and Transformations." In *Degeneration: The Dark Side of Progress,* ed. J. Edward Chamberlain and Sander L. Gilman, 263–89. New York: Columbia University Press, 1985.

Chamberlain, J. Edward, and Sander L. Gilman, eds. *Degeneration: The Dark Side of Progress.* New York: Columbia University Press, 1985.

Clarke, Bruce. *Allegories of Writing: The Subject of Metamorphosis.* Albany: State University of New York Press, 1995.

———. *Dora Marsden and Early Modernism: Gender, Individualism, Science.* Ann Arbor: University of Michigan Press, 1996.

———. "Resistance in Theory and the Physics of the Text." *New Orleans Review* 18, no. 1 (1991): 86–93.

Clarke, Bruce, and Linda Dalrymple Henderson, eds. *Modernism's Fourth Dimensions.* Forthcoming from the Pennsylvania State University Press.

Clifford, Gay. *The Transformations of Allegory.* Boston: Routledge, 1974.

Collier, John D. "Two Faces of Maxwell's Demon Reveal the Nature of Irreversibility." *Studies in the History and Philosophy of Science* 21, no. 2 (1990): 257–68.

Cooke, Leighton Brett. "Ancient and Modern Mathematics in Zamyatin's *We.*" In *Zamyatin's "We": A Collection of Critical Essays,* ed. Gary Kern, 149–67. Ann Arbor: Ardis, 1988.

———. "The Manuscript in Zamjatin's *We.*" *Russian Literature* 17 (1985): 367–88.

Cornford, Francis McDonald. *Plato's Cosmology: The "Timaeus" of Plato Translated with a Running Commentary.* London: Routledge and Kegan Paul, 1956.

Coveney, Peter, and Roger Highfield. *The Arrow of Time.* New York: Fawcett Columbine, 1990.

———. *Frontiers of Complexity.* New York: Fawcett Columbine, 1995.

Cowan, S. A. "The Crystalline Center of Zamyatin's *We.*" *Extrapolation* 29, no. 2 (1988): 160–78.

Crabtree, Adam. *From Mesmer to Freud: Magnetic Sleep and the Roots of Psychological Healing.* New Haven: Yale University Press, 1993.

Crawford, T. Hugh. "An Interview with Bruno Latour." *Configurations* 1, no. 2 (1993): 247–69.

Cushing, James T. *Philosophical Concepts in Physics: The Historical Relation between Philosophy and Scientific Theories.* New York: Cambridge University Press, 1998.

Dale, Peter Allan. *In Pursuit of a Scientific Culture: Science, Art, and Society in the Victorian Age.* Madison: University of Wisconsin Press, 1989.

Davies, Paul. *The Last Three Minutes.* London: Weidenfeld and Nicholson, 1994.

Davis, Erik. *Techgnosis: Myth, Magic, and Mysticism in the Age of Information.* New York: Harmony Books, 1998.

de Man, Paul. *Allegories of Reading: Figural Language in Rousseau, Nietzsche, Rilke, and Proust.* New Haven: Yale University Press, 1979.

———. "The Rhetoric of Temporality." In *Interpretation: Theory and Practice,* ed. Charles S. Singleton, 179–209. Baltimore: Johns Hopkins University Press, 1969.

de Vos, Luk, ed. *Just the Other Day: Essays on the Suture of the Future.* Antwerp: EXA, 1985.

Delavenay, Emile. *D. H. Lawrence and Edward Carpenter: A Study in Edwardian Transition.* New York: Taplinger, 1971.

Dobbs, Betty Jo Teeter, and Margaret C. Jacobs. *Newton and the Culture of Newtonianism.* Atlantic Highlands, N.J.: Humanities Press, 1995.

Duchamp, Marcel. *Marcel Duchamp: Notes.* Ed. and trans. Paul Matisse. Boston: G. K. Hall, 1983.

Eagleton, Terry. *Walter Benjamin; or, Towards a Revolutionary Criticism.* London: NLB, 1981.

Ehrenberg, Werner. "Maxwell's Demon." *Scientific American* 217 (1967): 103–10.

Einstein, Albert. "$E = MC^2$." In *Ideas and Opinions.* New York: Crown, 1954.

———. "Maxwell's Influence on the Evolution of the Idea of Physical Reality." In *Ideas and Opinions.*

———. *Relativity: The Special and General Theory.* Trans. Robert W. Lawson. 1920; New York: Crown, 1961.

Elliott, Robert S. *Electromagnetics: History, Theory, and Applications.* Piscataway, N.J.: IEEE Press, 1993.

Ellis, David. "Lawrence and the Biological Psyche." In *D. H. Lawrence: Centenary Essays,* ed. Mara Kalnins, 89–109. Bristol: Bristol Classical Press, 1986.

———. "Poetry and Science in the Psychology Books." In *D. H. Lawrence's Non-Fiction: Art, Thought, and Genre,* ed. David Ellis and Howard Mills, 67–97. New York: Cambridge University Press, 1988.

Ellis, David, and Howard Mills, eds. *D. H. Lawrence's Non-Fiction: Art, Thought, and Genre.* New York: Cambridge University Press, 1988.

Ellis, Edith. *Three Modern Seers.* New York: Mitchell Kennerly, 1910.

Eve, A. S., and C. H. Creasey. *Life and Work of John Tyndall.* London: Macmillan, 1945.

Everitt, C. W. F. "Maxwell's Scientific Creativity." In *Springs of Scientific Creativity: Essays on the Founders of Modern Science,* ed. Rutherford Aris, H. Ted Davis, and Roger H. Stuewer, 71–141. Minneapolis: University of Minnesota Press, 1983.

Fayter, Paul. "Strange New Worlds of Space and Time: Late Victorian Science and Science Fiction." In *Victorian Science in Context,* ed. Bernard Lightman, 256–80. Chicago: University of Chicago Press, 1997.

Fernihough, Anne. *D. H. Lawrence: Aesthetics and Ideology.* New York: Oxford University Press, 1993.

Ferry, Luc. *Homo Aestheticus: The Invention of Taste in the Democratic Age.* Trans. Robert De Loaiza. Chicago: University of Chicago Press, 1993.

Fineman, Joel. "The Structure of Allegorical Desire." In *The Subjectivity Effect in Western Literary Tradition: Essays toward the Release of Shakespeare's Will.* Cambridge: MIT Press, 1991.

Flammarion, Camille. *La Fin du Monde.* Paris: Ernest Flammarion, 1894.

———. *Omega: The Last Days of the World.* New York: Cosmopolitan Publishing, 1894; rpt. New York: Arno Press, 1975.

Fletcher, Angus. *Allegory: Theory of a Symbolic Mode.* Ithaca, N.Y.: Cornell University Press, 1964.

Foucault, Michel. *The Archaeology of Knowledge.* Trans. A. M. Sheridan Smith. New York: Pantheon Books, 1972.

Gauld, Alan. *A History of Hypnotism.* New York: Cambridge University Press, 1992.

Geduld, Harry M., ed. *The Definitive Time Machine.* Bloomington: Indiana University Press, 1987.

Genuth, Sara Schechner. *Comets, Popular Culture, and the Birth of Modern Cosmology.* Princeton: Princeton University Press, 1997.

Gieryn, Thomas F. *Cultural Boundaries of Science: Credibility on the Line.* Chicago: University of Chicago Press, 1999.

Gilbert, Elliot L. "'Upward, Not Northward': *Flatland* and the Quest for the New." *English Literature in Transition* 34, no. 4 (1991): 391–404.

Gillispie, Charles Coulston. *The Edge of Objectivity: An Essay in the History of Scientific Ideas.* Princeton: Princeton University Press, 1960.

Goldman, Martin. *The Demon in the Aether: The Story of James Clerk Maxwell.* Edinburgh: Paul Harris, 1983.

Grandville, J. J. *Fantastic Illustrations of Grandville: 266 Illustrations from "Un Autre Monde" and "Les Animaux."* New York: Dover, 1974.

Gregg, Richard A. "Two Adams and Eve in the Crystal Palace: Dostoevsky, the Bible, and *We.*" In *Zamyatin's "We": A Collection of Critical Essays,* ed. Gary Kern, 61–69. Ann Arbor: Ardis, 1988.

Hacking, Ian. "Probability and Determinism, 1650–1900." In *Companion to the History of Modern Science,* ed. R. C. Olby et al., 690–701. New York: Routledge, 1990.

Hakfoort, Casper. "The Historiography of Scientism: A Critical Review." *History of Science* 33 (1995): 375–95.

———. "Science Deified: Wilhelm Ostwald's Energeticist World-View and the History of Scientism." *Annals of Science* 49 (1992): 525–44.

Hall, Donald E., ed. *Muscular Christianity: Embodying the Victorian Age.* Cambridge: Cambridge University Press, 1995.

Haraway, Donna J. "A Cyborg Manifesto: Science, Technology, and Socialist-Feminism in the Late Twentieth Century." In *Simians, Cyborgs, and Women,* 149–81.

———. "Living Images: Conversations with Lynn Randolph." In *Millennial Myths:*

Painting by Lynn Randolph, ed. Marilyn A. Zeitlin, 23–33. Tempe: Arizona State University Press, 1997.

———. *Modest_Witness@Second_Millennium.FemaleMan©_Meets_OncoMouse™: Feminism and Technoscience*. New York: Routledge, 1997.

———. *Simians, Cyborgs, and Women: The Reinvention of Nature*. New York: Routledge, 1991.

Harman, P. M. "Edinburgh Philosophy and Cambridge Physics: The Natural Philosophy of James Clerk Maxwell." In *Wranglers and Physicists*, 202–24.

———. *Energy, Force, and Matter: The Conceptual Development of Nineteenth-Century Physics*. New York: Cambridge University Press, 1982.

———. *The Natural Philosophy of James Clerk Maxwell*. Cambridge: Cambridge University Press, 1998.

———, ed. *Wranglers and Physicists: Studies on Cambridge Physics in the Nineteenth Century*. Manchester: Manchester University Press, 1985.

Hawking, Stephen W. *A Brief History of Time*. New York: Bantam, 1988.

Hayles, N. Katherine. "The Ambivalent Approach: D. H. Lawrence and the New Physics." *Mosaic* 15, no. 3 (1982): 89–108.

———. *Chaos Bound: Orderly Disorder in Contemporary Literature and Science*. Ithaca, N.Y.: Cornell University Press, 1990.

———. *The Cosmic Web: Scientific Field Models and Literary Strategies in the Twentieth Century*. Ithaca, N.Y.: Cornell University Press, 1984.

———. *How We Became Posthuman: Virtual Bodies in Cybernetics, Literature, and Informatics*. Chicago: University of Chicago Press, 1999.

———. "'A Metaphor of God Knew How Many Parts': The Engine That Drives *The Crying of Lot 49*." In *New Essays on "The Crying of Lot 49,"* ed. Patrick O'Donnell, 97–125. New York: Cambridge University Press, 1991.

———, ed. *Chaos and Order: Complex Dynamics in Literature and Science*. Chicago: University of Chicago Press, 1991.

Heimann, P. M. "The *Unseen Universe:* Physics and the Philosophy of Nature in Victorian Britain." *British Journal for the History of Science* 21 (1972): 73–79.

Henderson, Linda Dalrymple. *Duchamp in Context: Science and Technology in the "Large Glass" and Related Works*. Princeton: Princeton University Press, 1998.

———. *The Fourth Dimension and Non-Euclidean Geometry in Modern Art*. Princeton: Princeton University Press, 1983.

———. "Mabel Dodge, Gertrude Stein, and Max Weber: A Four-Dimensional Trio." *Arts*, September 1982, 106–11.

———. "Modern Art and the Invisible: The Unseen Waves and Dimensions of Occultism and Science." In *Okkultismus und Avantgarde 1900–1915*. Frankfurt am Main: Schirn Kunsthalle Frankfurt, 1995.

———. "Mysticism as the 'Tie That Binds': The Case of Edward Carpenter and Modernism." *Art Journal* 46 (spring 1987): 29–37.

———. "Mysticism, Romanticism, and the Fourth Dimension." In *The Spiritual in Art: Abstract Painting, 1890–1985*. New York: Abbeville, 1986.

Hesse, Mary. "Models, Metaphors and Truth." In *From a Metaphorical Point of*

View: A Multidisciplinary Approach to the Cognitive Content of Metaphor, ed. Zdravko Radman, 351–72. Berlin: de Gruyter, 1995.

Hiebert, Erwin N. "The Uses and Abuses of Thermodynamics in Religion." *Daedalus* 95, no. 4 (1966): 1046–80.

Hinton, Charles Howard. *The Fourth Dimension.* 1904; rpt. New York: Arno Press, 1976.

———. *A New Era of Thought* (1888). London: Swan Sonnenschein, 1910.

———. *The Persian King.* In *Scientific Romances,* 1:33–128.

———. "A Picture of Our Universe." In *Scientific Romances,* 1:161–204.

———. "A Plane World." In *Scientific Romances,* 1:135–59.

———. "The Recognition of the Fourth Dimension." In *Speculations on the Fourth Dimension.*

———. *Scientific Romances.* 1st series. London: Swan Sonnenschein, 1886; rpt. 1st and 2d series, New York: Arno Press, 1976.

———. *Speculations on the Fourth Dimension: Selected Writings of Charles H. Hinton.* Ed. Rudolf Rucker. New York: Dover, 1980.

———. "What is the Fourth Dimension?" In *Scientific Romances,* 1:3–32.

Hollinger, Veronica. "Deconstructing the Time Machine." *Science-Fiction Studies* 14, no. 2 (1987): 201–21.

Hume, Kathryn. "Eat or Be Eaten: H. G. Wells's *Time Machine.*" *Philological Quarterly* 69, no. 2 (1990): 233–51.

Hunt, Bruce J. *The Maxwellians.* Ithaca, N.Y.: Cornell University Press, 1991.

Irigaray, Luce. *Speculum of the Other Woman.* Trans. Gillian C. Gill. Ithaca, N.Y.: Cornell University Press, 1985.

James, Frank A. J. L. "Thermodynamics and Sources of Solar Heat, 1846–1862." *British Journal for the History of Science* 15 (1982): 155–81.

Jay, Martin. "Modernism and the Specter of Psychologism." *Modernism/Modernity* 3, no. 2 (1996): 93–111.

Jeans, Sir James. *The Mysterious Universe.* Cambridge: Cambridge University Press, 1930.

Joravsky, David. *Soviet Marxism and Natural Science, 1917–1932.* New York: Columbia University Press, 1961.

Josephson, Paul R. *Totalitarian Science and Technology.* Atlantic Heights, N.J.: Humanities Press, 1996.

Kaku, Michio. *Hyperspace: A Scientific Odyssey through Parallel Universes, Time Warps, and the Tenth Dimension.* New York: Oxford University Press, 1994.

Kargon, Robert. "Model and Analogy in Victorian Science: Maxwell's Critique of the French Physicists." *Journal of the History of Ideas* 30 (1969): 423–36.

Kelley, Theresa M. *Reinventing Allegory.* Cambridge: Cambridge University Press, 1997.

Kern, Gary, ed. *Zamyatin's "We": A Collection of Critical Essays.* Ann Arbor: Ardis, 1988.

Kinkead-Weekes, Mark. *D. H. Lawrence: Triumph to Exile, 1912–1922.* New York: Cambridge University Press, 1996.

Kittler, Friedrich A. *Gramophone, Film, Typewriter*. Trans. Geoffrey Winthrop-Young and Michael Wutz. Stanford, Calif.: Stanford University Press, 1999.

Knapp, James F. *Literary Modernism and the Transformation of Work*. Evanston, Ill.: Northwestern University Press, 1988.

Knight, David. *The Age of Science: The Scientific World-View in the Nineteenth Century*. Oxford: Basil Blackwell, 1986.

Knott, Cargill Gilston. *Life and Scientific Work of Peter Guthrie Tait*. Cambridge: Cambridge University Press, 1911.

Kraut, Richard, ed. *Plato's "Republic": Critical Essays*. New York: Rowman and Littlefield, 1997.

Krips, Henry, J. E. McGuire, and Trevor Melia, eds. *Science, Reason, and Rhetoric*. Pittsburgh: University of Pittsburgh Press, 1995.

Landsberg, Peter T. "From Entropy to God?" In *Thermodynamics: History and Philosophy*, ed. K. Martinás, L. Ropolyi, and P. Szegedi, 379–403. London: World Scientific, 1991.

Latour, Bruno. *Aramis; or, The Love of Technology*. Trans. Catherine Porter. Cambridge: Harvard University Press, 1996.

———. *Science in Action: How to Follow Scientists and Engineers through Society*. Cambridge: Harvard University Press, 1987.

———. *We Have Never Been Modern*. Trans. Catherine Porter. Cambridge: Harvard University Press, 1993.

Lawrence, D. H. *Aaron's Rod*. New York: Penguin, 1976.

———. *Birds, Beasts, and Flowers!* Santa Rosa, Calif.: Black Sparrow Press, 1992.

———. *The Crown*. In *Reflections on the Death of a Porcupine and Other Essays*.

———. *D. H. Lawrence and New Mexico*. Ed. Keith Sagar. Salt Lake City: Gibbs M. Smith, 1982.

———. "Him with His Tail in His Mouth." In *Reflections on the Death of a Porcupine and Other Essays*.

———. "Indians and Entertainment." In *Morning in Mexico*.

———. "Introduction to These Paintings." In *Phoenix*.

———. *Lady Chatterley's Lover*. New York: Modern Library, 1959.

———. "Morality and the Novel." In *Study of Thomas Hardy and Other Essays*.

———. *Morning in Mexico*. Ed. Ross Parmenter. Salt Lake City: Gibbs M. Smith, 1982.

———. *Phoenix: The Posthumous Papers of D. H. Lawrence* (1936). Ed. Edward D. McDonald. New York: Viking, 1968.

———. *Psychoanalysis and the Unconscious* and *Fantasia of the Unconscious*. New York: Viking, 1976.

———. *The Rainbow*. New York: Penguin, 1976.

———. "Reflections on the Death of a Porcupine." In *Reflections on the Death of a Porcupine and Other Essays*.

———. *Reflections on the Death of a Porcupine and Other Essays*. Ed. Michael Herbert. New York: Cambridge University Press, 1988.

———. *Study of Thomas Hardy and Other Essays.* Ed. Bruce Steele. New York: Cambridge University Press, 1985.

———. *The Symbolic Meaning: The Uncollected Versions of "Studies in Classic American Literature."* New York: Viking, 1964.

———. "Why the Novel Matters." in *Study of Thomas Hardy and Other Essays.*

———. "The Wilful Woman." In *D. H. Lawrence and New Mexico,* ed. Keith Sagar, 6–9. Salt Lake City: Gibbs M. Smith, 1982.

———. *Women in Love.* New York: Penguin, 1989.

Leatherbarrow, W. J. "Einstein and the Art of Yevgeny Zamyatin." *MLR* 82, no. 1 (1987): 142–51.

Leff, Harvey S., and Andrew F. Rex, ed. *Maxwell's Demon: Entropy, Information, Computing.* Princeton: Princeton University Press, 1990.

Levine, George. "*Little Dorrit* and Three Kinds of Science." In *Dickens and Other Victorians,* ed. Joanne Shattock, 3–24. New York: St. Martin's, 1988.

———, ed. *Realism and Representation: Essays on the Problem of Realism in Relation to Science, Literature, and Culture.* Madison: University of Wisconsin Press, 1993.

Levine, George, and Alan Rauch, ed. *One Culture: Essays in Science and Literature.* Madison: University of Wisconsin Press, 1987.

Lodge, Sir Oliver. *Ether and Reality: A Series of Discourses on the Many Functions of the Ether of Space.* London: Hodder and Stoughton, 1925.

———. *My Philosophy, Representing My Views on the Many Functions of the Ether of Space.* London: E. Benn, 1933.

Luhan, Mabel Dodge. *Lorenzo in Taos.* London: Martin Secker, 1933.

———. *Movers and Shakers.* New York: Harcourt, Brace, 1936.

Luhmann, Niklas. *Essays on Self-Reference.* New York: Columbia University Press, 1990.

Madsen, Deborah L. *The Postmodernist Allegories of Thomas Pynchon.* New York: St. Martin's, 1991.

———. *Rereading Allegory: A Narrative Approach to Genre.* New York: St. Martin's, 1994.

Markley, Robert M. "Boundaries: Mathematics, Alienation, and the Metaphysics of Cyberspace." *Configurations* 3, no. 3 (1994): 485–507.

Marsden, Dora. "The Over-and-Above in Life." *Freewoman,* February 1, 1912, 201–2.

Martinás, K., L. Ropolyi, and P. Szegedi, eds. *Thermodynamics: History and Philosophy.* London: World Scientific, 1991.

Marx, Karl. *Karl Marx: Selected Writings.* Ed. David McLellan. Oxford: Oxford University Press, 1977.

Maxwell, James Clerk. "Address to the Mathematical and Physical Sections of the British Association" (1870). In *Scientific Papers,* 2:215–29.

———. "Are There Real Analogies in Nature?" In *The Life of James Clerk Maxwell,* by Lewis Campbell and William Garnett, 235–44. London: Macmillan, 1882.

———. "Ether." In *Scientific Papers,* 2:763–75.

———. "Hermann Ludwig Ferdinand Helmholtz." In *Scientific Papers,* 2:592–98.

———. *Matter and Motion.* 1876; rpt. New York: Dover, n.d.

———. *Maxwell on Heat and Statistical Mechanics: On "Avoiding All Personal Enquiries" of Molecules.* Ed. Elizabeth Garber, Stephen G. Brush, and C. W. F. Everitt. Bethlehem, Pa.: Lehigh University Press, 1995.

———. "On Faraday's Lines of Force." In *Scientific Papers,* 1:155–229.

———. "On Physical Lines of Force." In *Scientific Papers,* 1:451–513.

———. "*Paradoxical Philosophy.*" In *Scientific Papers,* 2:756–62.

———. "Report on Tait's Lecture on Force:—B.A., 1876." In *The Life of James Clerk Maxwell,* by Lewis Campbell and William Garnett, 646–48. London: Macmillan, 1882.

———. *The Scientific Letters and Papers of James Clerk Maxwell.* Ed. P. M. Harman. 2 vols. Cambridge: Cambridge University Press, 1990.

———. *The Scientific Papers of James Clerk Maxwell.* Ed. W. D. Niven. 2 vols. 1890; rpt. New York: Dover, n.d.

———. *Theory of Heat.* 9th ed. London: Longmans, Green, 1888.

———. "A Vision of a Wrangler, of a University, of Pedantry, and of Philosophy." In *The Life of James Clerk Maxwell,* by Lewis Campbell and William Garnett, 612–17. London: Macmillan, 1882.

Mazlish, Bruce. "A Triptych: Freud's *The Interpretation of Dreams,* Rider Haggard's *She,* and Bulwer-Lytton's *The Coming Race.*" *Comparative Studies in Society and History* 35, no. 4 (1993): 726–45.

McCrae, Murdo William, ed. *The Literature of Science: Perspectives on Popular Scientific Writings.* Athens: University of Georgia Press, 1993.

McHale, Brian. *Postmodernist Fiction.* New York: Methuen, 1987.

Mesmer, Franz Anton. *Mesmerism: A Translation of the Original Scientific and Medical Writings of F. A. Mesmer.* Trans. George Bloch. Los Altos, Calif.: William Kaufmann, 1980.

Morrison, Margaret. "A Study in Theory Unification: The Case of Maxwell's Electromagnetic Theory." *Studies in the History and Philosophy of Science* 23, no. 1 (1992): 103–45.

Moulthrop, Stuart. "Paragnosis, or the Story of Cyberspace." *Works and Days* 12, nos. 1–2 (1994): 33–55.

Moyer, Albert. *American Physics in Transition: A History of Conceptual Change in the Late Nineteenth Century.* Los Angeles: Tomash Publishers, 1986.

Myers, Greg. "Nineteenth-Century Popularizations of Thermodynamics and the Rhetoric of Social Prophecy." In *Energy and Entropy: Science and Culture in Victorian Britain,* ed. Patrick Brantlinger, 307–38. Bloomington: Indiana University Press, 1989.

Nersessian, Nancy J. "Kuhn and the Cognitive Revolution." In *Configurations* 6, no. 1 (1998): 87–120.

Newcomb, Simon. "Modern Mathematical Thought." *Nature,* February 1, 1894, 325–29.

Olby, R. C., et al., eds. *Companion to the History of Modern Science.* New York: Routledge, 1990.

Olson, Richard. *Scottish Physics and British Physicists.* Princeton: Princeton University Press, 1975.

Oppenheim, Janet. *The Other World: Spiritualism and Psychical Research in England, 1850–1914.* New York: Cambridge University Press, 1985.

Osietzki, Maria. "Rudolf Clausius: Entropy and Environment." In *Thermodynamics: History and Philosophy,* ed. K. Martinás, L. Ropolyi, and P. Szegedi, 57–62. London: World Scientific, 1991.

Ouspensky, P. D. *Tertium Organum: A Key to the Enigmas of the World.* Trans. Nicholas Bessaraboff and Claude Bragdon. Rochester, N.Y.: Manas Press, 1920.

Ovid. *The Metamorphoses.* Trans. Horace Gregory. New York: Viking, 1958.

Paradis, James, and Thomas Postlewait, eds. *Victorian Science and Victorian Values: Literary Perspectives.* 1981; New Brunswick, N.J.: Rutgers University Press, 1985.

Paulson, William R. *The Noise of Culture: Literary Texts in a World of Information.* Ithaca, N.Y.: Cornell University Press, 1988.

Paxson, James J. "The Allegory of Temporality and the Early Modern Calculus." *Configurations* 4, no. 1 (1996): 39–66.

———. *The Poetics of Personification.* New York: Cambridge University Press, 1994.

Penrose, Roger. *The Emperor's New Mind: Concerning Computers, Minds, and the Laws of Physics.* New York: Oxford University Press, 1990.

Peterfreund, Stuart, ed. *Literature and Science: Theory and Practice.* Boston: Northeastern University Press, 1990.

———. "The Re-Emergence of Energy in the Discourse of Literature and Science." *Annals of Scholarship: Metastudies of the Humanities and Social Sciences* 4 (1986–87): 22–53.

Philmus, Robert M. "The *Time Machine;* or, The Fourth Dimension as Prophecy." *PMLA* 84 (1969): 530–35.

Pitts, Mary Ellen. "Reflective Scientists and the Critique of Mechanistic Metaphor." In *The Literature of Science,* ed. McCrae, 249–72.

Plato. *The Republic.* Trans. Benjamin Jowett. New York: Modern Library, 1941.

Pollnitz, Christopher. "D. H. Lawrence's Quarrel with P. D. Ouspensky: Which Door to the Fourth Dimension?" In *Modernism's Fourth Dimensions,* ed. Bruce Clarke and Linda Dalrymple Henderson. Forthcoming from the Pennsylvania State University Press.

Preyer, Robert O. "The Romantic Tide Reaches Trinity: Notes on the Transmission and Diffusion of New Approaches to Traditional Studies at Cambridge, 1820–1840." In *Victorian Science and Victorian Values,* ed. Paradis and Postlewait, 39–68.

Prigogine, Ilya. *The End of Certainty: Time, Chaos, and the New Laws of Nature.* New York: Free Press, 1997.

Prigogine, Ilya, and Isabelle Stengers. *Order Out of Chaos: Man's New Dialogue with Nature.* New York: Bantam, 1984.

Pynchon, Thomas. *The Crying of Lot 49.* New York: Harper and Row, 1990.

Quilligan, Maureen. *The Language of Allegory: Defining the Genre.* Ithaca, N.Y.: Cornell University Press, 1979.

Rabinbach, Anson. *The Human Motor: Energy, Fatigue, and the Origins of Modernity.* New York: Basic Books, 1990.

Rhodes, Carolyn H. "Frederick Winslow Taylor's System of Scientific Management in Zamyatin's *We.*" *Journal of General Education* 28, no. 1 (1976): 31–42.

Risse, Guenter B. "Kant, Schelling, and the Early Search for Philosophical 'Science' of Medicine in Germany." *Journal of the History of Medicine and Allied Sciences* 27 (1972): 145–58.

Robbin, Tony. *Fourfield: Computers, Art, and the Fourth Dimension.* Boston: Little, Brown, 1992.

Rodwell, G. F. "On Space of Four Dimensions." *Nature,* May 1, 1873, 8–9.

Ropolyi, L. "Thermodynamic Elements in World Views—World View Elements in Thermodynamics." In *Thermodynamics: History and Philosophy,* ed. Martinás, Ropolyi, and Szegedi, 424–39.

Rothschuh, Karl E. *History of Physiology.* Trans. Guenter B. Risse. Huntington, N.Y.: R. E. Krieger, 1973.

Rothstein, Jerome. "Physical Demonology." *Methodos* 42 (1959): 99–121.

Rucker, Rudolf. *Geometry, Relativity, and the Fourth Dimension.* New York: Dover, 1977.

Rudnick, Lois Palken. *Mabel Dodge Luhan: New Woman, New Worlds.* Albuquerque: University of New Mexico Press, 1984.

Sawyer, Paul L. "Ruskin and Tyndall: The Poetry of Matter and the Poetry of Spirit." In *Victorian Science and Victorian Values,* ed. Paradis and Postlewait, 217–46.

Scafella, Frank. "From Reason to Intelligence: Wells's White Sphinx as Chronotope of Nineteenth Century Science." *Cahiers Victoriens et Edouardiens* 46 (1997): 181–89.

———. "The White Sphinx and *The Time Machine.*" *Science Fiction Studies* 8, no. 3 (1981): 255–65.

Scherr, Barry J. *D. H. Lawrence's Response to Plato: A Bloomian Interpretation.* New York: Peter Lang, 1996.

Schweber, Silvan S. "Demons, Angels, and Probability: Some Aspects of British Science in the Nineteenth Century." In *Physics as Natural Philosophy,* ed. Abner Shimony and Herman Feshbach, 332–33. Cambridge: MIT Press, 1982.

Serres, Michel. *Hermes: Literature, Science, Philosophy.* Ed. Josué V. Harari and David F. Bell. Baltimore: Johns Hopkins University Press, 1982.

———. *Angels: A Modern Myth.* New York: Flammarion, 1995.

Serres, Michel, with Bruno Latour. *Conversations on Science, Culture, and Time.* Trans. Roxanne Lapidus. Ann Arbor: University of Michigan Press, 1995.

Shankman, Steven, ed. *Plato and Postmodernism.* Glenside, Pa.: Aldine Press, 1994.

Shimony, Abner, and Herman Feshbach, eds. *Physics as Natural Philosophy.* Cambridge: MIT Press, 1982.

Sicher, Efraim. "The Last Utopia: Entropy and Revolution in the Poetics of Evgeny Zamjatin." *History of European Ideas* 13, no. 3 (1991): 225–37.

Siegel, Daniel M. "Thomson, Maxwell, and the Universal Ether in Victorian Physics." In *Conceptions of Ether: Studies in the History of Ether Theories, 1740–1900,* ed. G. N. Cantor and M. J. S. Hodge, 239–68. New York: Cambridge University Press, 1981.

———. "Mechanical Image and Reality in Maxwell's Electromagnetic Theory." In *Wranglers and Physicists: Studies on Cambridge Physics in the Nineteenth Century,* ed. P. M. Harman, 180–201. Manchester: Manchester University Press, 1985.

Sklar, Lawrence. *Space, Time, and Spacetime.* Berkeley and Los Angeles: University of California Press, 1974.

Slade, Joseph W., and Judith Yaross Lee, eds. *Beyond the Two Cultures: Essays on Science, Technology, and Literature.* Ames: Iowa State University Press, 1990.

Smee, Alfred. *Elements of Electro-Biology, or, The Voltaic Mechanism of Man; of Electro-Pathology, especially of the Nervous system; and of Electro-Therapeutics.* London: Longman, Brown, 1849.

Smith, Crosbie. "Natural Philosophy and Thermodynamics: William Thomson and 'The Dynamical Theory of Heat.'" *British Journal for the History of Science* 9 (1976): 293–319.

———. *The Science of Energy: A Cultural History of Energy Physics in Victorian Britain.* Chicago: University of Chicago Press, 1998.

Smith, Crosbie, and M. Norton Wise. *Energy and Empire: A Biographical Study of Lord Kelvin.* New York: Cambridge University Press, 1989.

Smith, Jonathan. *Fact and Feeling: Baconian Science and the Nineteenth-Century Literary Imagination.* Madison: University of Wisconsin Press, 1994.

Spencer, Herbert to John Tyndall, no date (c. 1858). Royal Institution of Great Britain. Tyndall Correspondence, vol. 3, p. 1175 (19/D1.90).

Stallo, J. B. *The Concepts and Theories of Modern Physics.* London: Kegan Paul, Trench, 1882.

Stewart, Balfour. *The Conservation of Energy, Being an Elementary Treatise on Energy and its Laws.* 5th ed. London: Kegan Paul, 1881.

Stewart, Balfour, and P. G. Tait, *The Unseen Universe or Physical Speculations on a Future State.* New York: Macmillan, 1875.

Stites, Richard. *Revolutionary Dreams: Utopian Vision and Experimental Life in the Russian Revolution.* New York: Oxford University Press, 1989.

Suvin, Darko. "A Grammar of Form and a Criticism of Fact: *The Time Machine* as a Structural Model for Science Fiction." In *H. G. Wells and Modern Science Fiction,* ed. Darko Suvin and Robert M. Philmus, 90–115. Lewisburg, Pa.: Bucknell University Press, 1976.

———. *Metamorphoses of Science Fiction: On the Poetics and History of a Literary Genre.* New Haven: Yale University Press, 1979.

———. "Science-Fiction: Metaphor, Parable and Chronotope." In *Just the Other*

Day: Essays on the Suture of the Future, ed. Luk de Vos, 81–99. Antwerp: EXA, 1985.

Suvin, Darko, and Robert M. Philmus, eds. *H. G. Wells and Modern Science Fiction.* Lewisburg, Pa.: Bucknell University Press, 1976.

Tait, Peter Guthrie. *Sketch of Thermodynamics.* Edinburgh: Edmonston and Douglas, 1868.

Tait, Peter Guthrie to James Clerk Maxwell. December 6, 1867. Cambridge University Library, Add 7655/Ia/4.

Tatar, Maria M. *Spellbound: Studies on Mesmerism and Literature.* Princeton: Princeton University Press, 1978.

Tedlock, E. W., Jr. "D. H. Lawrence's Annotations of Ouspensky's *Tertium Organum.*" *Texas Studies in Literature and Language* 2, no. 2 (1960): 206–18.

Teskey, Gordon. *Allegory and Violence.* Ithaca, N.Y.: Cornell University Press, 1996.

Theobald, D. W. *The Concept of Energy.* London: E. and F. N. Spon, 1966.

Thompson, S. P. *The Life of William Thomson, Baron Kelvin of Largs.* 2 vols. London, 1910.

Thomson, William. "The Kinetic Theory of the Dissipation of Energy" (1874). In *Mathematical and Physical Papers,* ed. Larmor, 5:11–20.

———. *Mathematical and Physical Papers.* Ed. Sir Joseph Larmor. 5 vols. Cambridge: Cambridge University Press, 1911.

———. "On a Universal Tendency in Nature to the Dissipation of Mechanical Energy." In *Mathematical and Physical Papers,* 1:511–14.

———. "On the Age of the Sun's Heat." *Macmillan's,* March 1862, 388–93.

———. "The Sorting Demon of Maxwell" (1879). In *Mathematical and Physical Papers,* ed. Larmor, 5:21–23.

Tindall, William York. *D. H. Lawrence and Susan His Cow.* New York: Columbia University Press, 1939.

Trembley, Anthony, and Archie Henderson. "Modernism and the Occult: A Selected Bibliography." In *Literary Modernism and the Occult Tradition,* ed. Leon Surette and Demetres Tryphonopoulos, 1–16. Orono, Maine: National Poetry Foundation, 1996.

Troward, Thomas. *The Edinburgh Lectures on Mental Science.* New York: Robert M. McBride, 1909.

Tyndall, John. "The Constitution of Nature" (1865). In *Fragments of Science,* 1:3–27.

———. *Fragments of Science: A Series of Detached Essays, Addresses, and Reviews.* 2 vols. New York: D. Appleton, 1915.

———. *Heat: A Mode of Motion.* 6th ed. New York: D. Appleton, 1915.

———. "Lecture X." In *Heat.*

———. "On Radiant Heat in Relation to the Colour and Chemical Constitution of Bodies" (1866). In *Fragments of Science,* 1:74–95.

———. "Radiation" (1865). In *Fragments of Science,* 1:28–73.

———. "Scientific Use of the Imagination" (1870). In *Fragments of Science,* 2:101–34.

Underwood, Ted. "Productivism and the Vogue for 'Energy" in Late Eighteenth-Century Britain." *Studies in Romanticism* 34, no. 1 (1995): 103–25.

Upward, Allen. *The New Word.* New York: Mitchell Kennerly, 1910.

von Neumann, John. "Method in the Physical Sciences." In *Collected Works,* ed. A. H. Taub, 491–98. New York: Macmillan, 1963.

Washington, Peter. *Madame Blavatsky's Baboon: A History of the Mystics, Mediums, and Misfits Who Brought Spiritualism to America.* New York: Schocken, 1995.

Webb, James. Introduction to *Scientific Romances,* by Charles Howard Hinton. New York: Arno Press, 1976.

Wells, H. G. "The Star." In *The Complete Short Stories of H. G. Wells.* New York: St. Martin's, 1970.

———. *The Time Machine.* New York: Random House, 1931.

———. *The Time Machine.* Ed. Michael Moorcock. Rutland, Vt.: Charles E. Tuttle, 1993.

Wilczek, Frank. "The Persistence of Ether." *Physics Today,* January 1999, 11–13.

Williams, Linda Ruth. *Sex in the Head: Visions of Femininity and Film in D. H. Lawrence.* Detroit: Wayne State University Press, 1993.

Wilson, David. "The Educational Matrix: Physics Education at Early-Victorian Cambridge, Edinburgh, and Glasgow Universities." In *Wranglers and Physicists: Studies on Cambridge Physics in the Nineteenth Century,* ed. P. M. Harman, 12–48. Manchester: Manchester University Press, 1985.

———. "Kelvin's Scientific Realism: The Theological Context." *Philosophical Journal* 11 (1974): 41–60.

Winter, Alison. *Mesmerized: Powers of Mind in Victorian Britain.* Chicago: University of Chicago Press, 1998.

Wutz, Michael. "The Energetics of Tarr: The Vortex-Machine Kreisler." *Modern Fiction Studies* 38, no. 4 (1992): 845–69.

Young, Richard O. "'Where Even the Trees Come and Go': D. H. Lawrence and the Fourth Dimension." *D. H. Lawrence Review* 13, no. 1 (1980): 30–44.

Zamyatin, Yevgeny. "H. G. Wells." In *A Soviet Heretic.*

———. "The New Russian Prose." In *A Soviet Heretic.*

———. "On Literature, Revolution, Entropy, and Other Matters." In *A Soviet Heretic.*

———. "On Synthetism." In *A Soviet Heretic.*

———. *A Soviet Heretic: Essays by Yevgeny Zamyatin.* Ed. and trans. Mirra Ginsburg. Chicago: University of Chicago Press, 1970.

———. *We.* Trans. Clarence Brown. New York: Penguin, 1993.

Zapf, Hubert. "Taylorism in D. H. Lawrence's *Women in Love.*" *D. H. Lawrence Review* 15, nos. 1–2 (1982): 129–39.

Zencey, Eric. "Entropy as Root-Metaphor." In *Beyond the Two Cultures: Essays on Science, Technology, and Literature,* ed. Joseph W. Slade and Judith Yaross Lee, 185–200. Ames: Iowa State University Press, 1990.

Index